FAILING PROJECTS IN THE MODERN WORLD

WHY DO LARGE PROJECTS FAIL?

DR. MARCO GARCIA

First Published 2023 by Marco Garcia

Copyright © 2023 Marco Garcia

ISBN 978-0-7961-3965-8 (Print)

ISBN 978-0-7961-3966-5 (eBook)

Cover and Interior crafted with love by the team at www.myebook.online

All rights reserved.

No part of this publication may be reproduced, stored in or introduced into a retrieval system, or transmitted, in any form, or by any means electronic, mechanical, photocopying, recording or otherwise, without the prior permission of the publisher. Any person who commits any unauthorised act in relation to this publication may be liable to criminal prosecution and civil claims for damages.

ABOUT THIS BOOK

Are your construction projects failing?

Many construction projects fail today, not because of poor project planning, but because of poor project management and inefficient risk monitoring and control. Successful construction project managers efficiently manage the numerous environmental, financial, socio-economic and technical risks that impede the workflow of the construction project, its scope, and its time and cost planning.

This book explores the factors that lead to efficient and inefficient risk management in the construction sector: It identifies the key risks associated with construction projects, the role and significance of risk management, the causes of project failure, and the strategies and approaches to risk management that minimize failure.

Its information is based on a review of the risks and risk management practices in the UK construction sector and a 2022 survey of 100 UK project managers' views on factors that contribute to the failure of projects, such as scheduling-oriented challenges, resource scarcities and technological malfunctions, as well as methodologies and tools that contribute to the success of projects, such as the critical path method and Gantt chart technique, artificial intelligence and agile methodologies, including the scrum technology.

ABOUT THE AUTHOR

Marco Garcia is married with two daughters. Born in South Africa of a Portuguese father and a British mother, he has a PhD in project management and 28 years of experience in the construction industry, working on small to large projects in the fields of power generation, oil and gas, and green energy, in Africa, Brazil, Hong Kong, Iraq and Mexico.

Marco was part of three simultaneous mega projects in Egypt and, in 2019, they received the Project of the Year Award from *Power Magazine* (US).

CONTENTS

About this book iii

About the author... iv

List of Abbreviations xv

CHAPTER 1: WHY IS IT IMPORTANT HOW LARGE PROJECTS ARE MANAGED?..1

What happens when a project is managed? 3

What is different about managing a large project? 5

Classic project management theory..6
- *Planning*...6
- *Control*...6
- *Execution*..7

Large project management theory 8
- *Planning and scheduling*..8
- *Project readiness*...9
- *Cost control*...9
- *New knowledge for new projects* 9
- *High-quality results* 10
- *Continuous oversight* 10
- *Teamwork and collaboration* 10
- *Adaptive management practices*..11

How are large projects managed? 11

Goals and planning around cost, quality and time.................12
Control..13
Communication 14
Resource management...15
Long-term project management...16

How are large projects managed in the UK? 16

    Methodologies, definitions, and theories 18

    Classic project management 19

    Agile project management 20

    Critical Chain Project Management 22

    Project Evaluation and Review Technique 22

    Critical Path Technique 23

    Kanban 24

    Lean software development 24

    Methodology for PRojects IN Controlled Environments 25

    Scrum 25

    Waterfall 27

    Work breakdown structure 27

Conclusions about project management of large projects 28

    Chapter summary 29

CHAPTER 2: WHY IS EFFECTIVE RISK MANAGEMENT IMPORTANT IN LARGE PROJECTS? 31

Types of risk 32

    Known and unknown risks 33

    Fundamental, particular, pure and speculative risks 34

    Cost, performance and scheduling risks 34

    Factor-related risks 35

        Communication-related risks *36*

        Culture-related risks *36*

        Management-related risks *36*

        Scope-related risks *37*

        Quality-related risks *38*

        Finance-related risks *38*

        Worker-related risks *38*

What is effective risk management? 39

Iterative risk management process..42
    Risk identification..44
    Risk assessment..45
    Risk response...47
    Risk monitoring...48

Risk management tools and techniques 48
    Project risk management..48
    Root cause analysis..49
    Risk avoidance, transfer, mitigation and acceptance.........50
    Risk registers..50
    Total quality management..51
    Just-in-time..52
    Strategies..53

Consequences of not implementing
effective risk management..54

Conclusions about the importance of risk management............55

CHAPTER 3: WHY HAVE CONSTRUCTION PROJECTS FAILED IN THE UNITED KINGDOM? 57

Why I investigated the problems..57

What were the problems? 58

How other people saw the problems..................................59
    Project management concepts..59
    Project management risks...61
        Communication-related risks......................................61
        Resource-based risks..62
        Scheduling risks..63
        Human resource risks...63
    Project management theories...65
        Scrum methodology..65
        Supervision..65
        Scope management..66

Contemporary construction project failures 66
    Project leadership *67*
    Budgeting *67*
    Performance management and control *68*
    Artificial intelligence *68*
    Collaboration *68*
    Communication *69*
    Conflict management *70*

Factors that make projects unsuccessful 71
    Budget overruns *72*
    Resource mismanagement *72*
    Scheduling mismanagement *73*
    Technological malfunctions *73*

Possible measures to eliminate the
causes of project failure 74
    Building information modeling *74*
    Procurement models *75*
    Research and development *75*
    The theory of inventive problem-solving *77*
    Workforce-related measures *77*
    Waste reduction measures *78*

Conclusions from the literature review 78

Conclusions from my research 78
    Resource scarcity *79*
    Scheduling-oriented challenges *81*
    Technological malfunctions *83*

CHAPTER 4: WHAT SHOULD PROJECT MANAGERS DO? 85

What is project team management? 86

Why is teamwork important? 86

Teamwork principles 87

Team management weaknesses 87

Teams and working groups 88

Team building 88

Team performance 90

Homogeneous and heterogeneous teams 91

Belbin's team roles 91

Power sources 92

Different styles of group and individual management 93

How to build a productive team 94
    Project team meetings 95
    Communication and information 96
    Motivation and stimulation 96

Competence to manage the project team 97

Technical and professional competencies 99

Certification by the International
Project Management Association 100

Virtual teams course completion and evaluation 104

Project Management Institute methodologies
for managing project teams 104

PRINCE2 methodologies for managing project teams 105

Situational management and leadership 106

Teams in the organizational context 106

Managing the project team 106

Aligning organizational goals and project goals 107

Delegating and transmitting information 108

Risks of team management 108

Team mood and conflict resolution 108

Multicultural teams and diversity management 110

Conclusion 111

LIST OF REFERENCES 113

APPENDIX A: HOW I INVESTIGATED THE PROBLEMS IN THE UK CONSTRUCTION INDUSTRY 145

How I chose my research methodology 146

    Positivism 146
        A hypothetico-deductive model 146
        Objectivity and dualism 147

    A deductive approach 148

    A descriptive design 148

    A quantitative survey questionnaire 149

    A random sampling approach 149

    A pilot study 150
    The actual survey 150
    Quantitative analysis 150
    The variables 150

How I broke the problem down 151
    Ethical considerations 151
    Limitations 152

What I found out 152
    Inferential statistics 152
        Experience 153
        Key motive of project management
        in the construction industry 154
        Project failures are increasing 155
        Critical risks in project management 156
        Common scheduling challenges 157
        Critical path method and Gantt chart techniques 158
        Resource scarcity or inadequate
        management of resources 159
        Factors that influence the failure of
        construction projects in the UK 160
        Technology, project objectives,
        and access to information 162
        Tools and measures 163
        AI technology 164
        Agile methodology 165
        Scrum methodology 167
    Descriptive statistics 168
        Risks related to project management 169
        Scheduling-oriented challenges 170
        Technological malfunction 170
        Artificial intelligence 170
        Agile methodology 171
        Overall conclusions 171
    Regression analysis: what the statistics mean 172

APPENDIX B: SURVEY QUESTIONNAIRE 177

LIST OF TABLES

Table 1: Team building phases 89

Table 2: Belbin's team roles 91

Table 3: Elements of International Project Management Association certification 101

Table 4: Descriptive statistics 168

Table 5: Model summary 173

Table 6: Analysis of variance 174

Table 7: Coefficients 175

LIST OF FIGURES

Figure 1: Classification of risk (probability and impact) 34

Figure 2: Iterative risk management process57 44

Figure 3: Probability and impact matrix 46

Figure 4: Risk ratings matrix 46

Figure 5: Experience in the construction industry 153

Figure 6: Project management helps fulfill agreed goals 154

Figure 7: The incidence of project failures has increased in the last few years 155

Figure 8: Prominent risks in managing UK construction projects 156

Figure 9: Scheduling-related challenges make construction projects fail 157

Figure 10: The Critical Path Technique and Gantt charts in optimal scheduling 158

Figure 11: A scarcity of resources can result in a failed project 160

Figure 12: Key factors that cause the failure of construction projects in the UK 161

Figure 13: Technological malfunctioning results in project failures 162

Figure 14: Measures and tools that support effective and efficient project completion 163

Figure 15: Importance of AI technology in the UK construction sector 165

Figure 16: Importance of agile methodologies in effective change and completion 166

Figure 17: Scrum technology provides flexibility and effective task control 167

LIST OF ABBREVIATIONS

AI	Artificial Intelligence
ANOVA	Analysis of Variance
APM	Association for Project Management
BBC	British Broadcasting Corporation
BIM	Building Information Modeling
CAD	Computer-Aided Design
CCPM	Critical Chain Project Management
CPM	Critical Path Method
DSDM	Dynamic Systems Development Method
GDP	Gross Domestic Product
HAZOP	Hazard and Operability and Analysis
HSE	Health and Safety Executive
IPMA	International Project Management Association
JIT	Just-In-Time
PERT	Program Evaluation and Review Technique
PMBOK	Project Management Body of Knowledge

PMI	Project Management Institute
PRINCE2	PRojects IN Controlled Environments
PRM	Project Risk Management
RCA	Root Cause Analysis
SPSS	Statistical Package for the Social Sciences
TQM	Total Quality Management
TRIZ	Theory of Inventive Problem-solving (Russian acronym)
UK	United Kingdom
WBS	Work Breakdown Structure

CHAPTER 1

WHY IS IT IMPORTANT HOW LARGE PROJECTS ARE MANAGED?

> "
> Only 35% of projects today are completed successfully and on time [1]
> "

The Project Management Institute (PMI) defines a *project* as "any temporary endeavor with a definite beginning and end".[2] In layman's terms, a project is a set of tasks that must be completed within a defined timeline to accomplish a specific set of goals.

The success or failure of projects, small or large, significantly depends on the effectiveness of the management and execution of the various project-related processes and activities. *Project management* refers to the application of experience, knowledge, methods, processes and skills to accomplish the project goals set by defined parameters of cost, quality and time.

The scope for chaos in organizations expands when the projects they handle become more multifaceted. The multiple facets demand adequate planning, and this is what makes project management relevant. However, project management is changing substantially, and it is important to keep up with the changes in order to reap the advantages of project management.

The final deliverables or project outcomes in projects constrain the predetermined budget and timescale. Project managers must therefore plan, staff, direct, organize and control efficiently while working on the project because poor management of project activities directly affects the project outcomes.[3] But managing a project, particularly a large project, is difficult because the manager has to assess the cultural aspects of the team members, stakeholder participation and communication, identify potential risks, and handle contingent situations. Project managers must have the skills to tackle edgy issues related to overbudgeting, delays, scope creep and problematic team members, and these issues tend to magnify with the scale and size of the project.[4] Long-lasting projects pose management difficulties mainly due to the constant changes taking place in the market and overall business environment because these changes necessitate adjustments to all project elements.

Project management is a sound business tool that has several advantages for businesses of different sizes. It offers proper guidelines, techniques and repeatable processes to help manage the work and the individuals involved in a large project. Project management tactics can stimulate the chances of success, and aid in delivering projects continuously, efficiently, within budget and on time.[5]

Large projects are the most difficult to manage because there are high costs, multiple stakeholders and numerous activities associated with them where poor management aggravates issues resulting from unclear objectives and inadequate business focus.[6,7] Moreover, the complexities of a large-scale project demand particular attention to planning the project, selecting team members, delivering solutions, and maintaining a skilled and high-performing team. However, clear direction and leadership ensure the success of projects and the attainment of deliverables.[4]

Evidence from recent large-scale infrastructure projects in the UK suggests that project governance and success also depend on clients' permanence because permanent clients regularly undertake various projects, whereas a temporary client tends to disappear once the project is completed.[6,7]

This book evaluates the importance of management in 100 large-scale UK projects, explores the project management techniques that are widely used for executing such projects, and discusses the relevance of numerous books and articles on the subject.

WHAT HAPPENS WHEN A PROJECT IS MANAGED?

Effective project management helps decrease project costs by eradicating risks, improving efficacy and optimizing resources.[8] According to Yang et al.,[7] a project's success has five dimensions: efficacy, impact on the team, impact on customers, business outcomes and direct outcomes for the involved organizations. To be successful, the project also requires preparation for achieving time and budget goals, attaining customer satisfaction, fulfilling client requirements, and maintaining the morale and skill of project team members. The degree of the manager's success in completing projects is measured by criteria such as profits, sales, service quality, and whether new markets, technology or capabilities were developed. The conditions for successfully managing project processes are careful project planning, effective governance, defined goals, competent project teams, and commitment to success throughout its lifecycle.[7]

All projects have start dates and end dates that mark their lifecycle, and there are commissioning, planning and execution phases between these dates. The activities of planning, executing, monitoring, team management, and making necessary changes during the progress of the project all support the lifecycle activities of the project. How these are managed—the methodology—is unique to each project. For instance, Ericsson devised a unique in-house management process for defining the activities needed to develop mobile communication systems.[8]

Theoretical project management methodologies are commonly used in real-life practice by a wide range of large and small projects, where they have ensured better management of the projects and fostered project success.

WHAT IS DIFFERENT ABOUT MANAGING A LARGE PROJECT?

Large projects have more varied activities, more team members and more (different) stakeholders than smaller projects so they need exceptional levels of management and control to ensure that the desired project outcomes are achieved. Without proper planning, team leadership and management in the form of resource allocation and collaboration between different parties, it is not possible to meet project goals and successfully complete a large and complex project.[6]

Large projects and their complexities require project managers to direct their attention to planning the project, developing and delivering solutions, choosing team members, and maintaining consistent team performance over a lengthy period. Large projects need effective control and management of project activities, and they are more susceptible to failure than small projects due to their elevated levels of uncertainties and risks. To avoid failure, the theory of management of large projects can be encapsulated in three considerations: strong foundations, a semi-permeable boundary, and a transformational process view.[9]

The classic project management theory primarily focuses on the transformational processes that occur in discrete activities where one output is input to another activity—but that focus is not appropriate for large projects where project performance can be negatively affected by the inappropriate precedence and coupling of activities because these flows are not static in large projects.[9]

Before we consider further why and how large projects must be managed differently, let us look at the classic project management theory.

Classic project management theory

The two broad schools of thought in classic project management theory concerned administrative theories and scientific management. The key principles of the theory included planning and forecasting, organizing, commanding, coordinating and governing. Further, applying the three core management theories of planning, control and execution, which are part of project management, had a considerable impact on the outcomes or results.

Planning

Decisions are mainly taken in the planning phase. Changes needed to be incorporated into the planning process in response to modern market trends, and corrections were needed to control processes based on performance data.[10]

Control

The theory of control is a pivotal part of project management, where controlling contains two sub-processes: overall change control and performance reporting. Corrections are suggested for the execution phase, and changes are suggested to the plans from the planning phase in response to performance reporting while managing the project.

Users of the classic theory recognized that a project manager or team leader requires constant support from employees in achieving defined goals and providing deliverables. The key principles were therefore complemented by human relations theory and behavioral theory centered on the people facets of management and organizations. This was because management is a dynamic and ongoing procedure, and human capital is an active participant in project execution.

Execution

The concept of execution in the theory of management contains two constituents: communication or authorization, and the decision to select a particular task from a set of tasks. Execution mainly includes task allocation, communication about progress, and the need for improvement or changes.

Traditional project management practices focused on control, execution and planning are necessarily applied to managing different types of projects, but these management practices are not adequate for the successful completion of contemporary projects, whether they are small or large. This is because traditional project management practices are not compatible with the current business environment, which is characterized by continuous change, varied market trends and modern technologies.[11] For this reason, new project management theories were developed, which are more adaptable in the face of continuous change, varied market trends and modern technologies, and some of which are more appropriate for large projects.

Large project management theory

Managing a large project requires adaptive practices and effective project leadership. Effective project leadership contributes to the successful management of large and complex projects because project managers tend to practice situational leadership, and modify and adapt their approach or strategy to accommodate the inevitable changes.

Large projects burden the project team members because there are many deliverables, as well as quality and cost criteria, that hinder project success. Employee burnout and team fatigue result in complex human interactions, and generally create unavoidable staff turnover that is difficult to predict, thereby slowing down the pace of the project, and leading to delays or time overruns.[11,12]

Managers of large projects need to have a plan and a schedule that consider project readiness, the project lifecycle, cost controls, how to build new knowledge for new projects, and how to obtain high-quality results. They must maintain continuous oversight, foster teamwork and collaboration, and apply adaptive management practices. All these are enlarged on below.

Planning and scheduling

Consideration of appropriate management techniques in projects helps establish a proper management plan and project schedule. Sticking by the project schedule articulated in the management plan inculcates the discipline needed to avoid delays, and a predefined process via the project lifecycle provides a clear path to the project's completion, thereby increasing the likelihood of success.[13]

Project readiness

Project managers cannot initiate a project on the presupposition that the owner's firm is ready to execute the project because presuppositions can lead to incorrect conclusions. All presuppositions must, therefore, be verified before making any decisions or assumptions about project readiness. It is for this reason that verification of project readiness is a vital first step in the process of project initiation to ensure that the project is ready for development and implementation as planned. The verification should cover the thoroughness of the planning process, the adequacy and training of project personnel, and the availability and reserve of project resources, support services and systems.

Cost control

Keeping control of the costs is a major feature of successfully accomplished projects. Based on their scope, certain projects are likely to incur enormous costs. This makes it important to keep track of the project budget and integrate project management strategies to relieve the risks concerned with budget overruns.

New knowledge for new projects

Effective management is regarded as inseparable from the process of building the new knowledge required to embark on new projects.[14] When business entities take off with new projects, they gain more experience with time. Focusing on the management aspect provides a knowledge asset to the organization and helps build knowledge and competitiveness as well as experience.

High-quality results

Almost every entity dealing with projects emphasizes the importance of assuring high-quality results. Project management helps identify, control and optimally manage the standards that lead to the high-quality service and products that provide client satisfaction.[15]

Continuous oversight

Continuous oversight is regarded as vital in almost every project, regardless of the project's nature, type and size; and the management of projects ensures this constant oversight—or control. Project management methods ensure that organizations have sufficient control over ongoing projects to affirm that the projects are on the right track and within the established budget. Management of projects aids in delivering outputs for maximizing the benefits of an organization by providing clear objectives and focal points, guaranteeing quality cost management and adequate risk management.[16]

Teamwork and collaboration

Proper management must strive for effective teamwork because it helps ensure that the people in the project team collaborate to capture the benefits of knowledge sharing and an expanded skill set. Good project management also helps managers motivate team members to collaborate to meet the key purpose of the project.[17]

Adaptive management practices

Managing large projects requires a completely new outlook on the part of both project leader and team members to ensure that risks or uncertainties are minimized, and the probability of project success is increased. In order to plan and structure large projects, traditional practices need to be combined with adaptive management practices. Along with this combination, reliable and systematic approaches need to be deployed in addition to progressive elaboration for estimating the accuracy of project stages. Strict cost and time management enhance the credibility of project outcomes and reduce the gap between actual outcomes and desired outcomes while managing complex and lengthy projects. Stage-gate management accompanied by rigorous risk management facilitates continuous improvement and also enables project managers to predict potential challenges and exploit new opportunities.[18]

HOW ARE LARGE PROJECTS MANAGED?

Traditional project management offers a combination of tools, methods and processes to manage projects on time, within a specified budget and to the needful specifications. These widely used processes are valuable for determining the knowledge bodies and detailed procedures needed to establish and execute projects. However, traditional management has been criticized in the last few years for neglecting the consideration of strategic decisions and techniques, undertaken during the front-end planning phase of projects, that play a crucial role in articulating the key requirements of a project and ascertaining whether the goal is accomplished.

Many projects in the UK, mainly infrastructure projects, now realize the need for management solutions that are developed for dealing with the particular challenges of planning and executing large, complex projects.[19]

Goals and planning around cost, quality and time

A clearly outlined project goal and effective steps to attain that goal are assessed by cost, quality and time—the three constraints of management in projects—also termed the *iron triangle*. This is why the main role of the project manager is to complete the project by the specified time, within the defined budget and in line with clear specifications.

Through adequate management of projects, trade-offs between these three constraints are frequently made to accomplish the project's aim. For instance, although there were fixed times to efficiently complete the construction of the London 2012 Olympics infrastructure and various venues, the quality and budget could be altered to attain the goal.[20] If the time schedule is reduced, additional resources and a sufficient budget might be essential to deliver a project in relatively less time.[19] On the other hand, if the budget is reduced, the time schedule often has to be expanded. For instance, the Crossrail project in London was ordered to reduce the entire budget by £1 billion to meet the requirements defined in the government's comprehensive *Spending Review 2010*. To attain this refined constraint, the stations and tunnels were developed sequentially instead of in parallel, and the completion date of the project was postponed.[21]

Thus, appropriate management of goal setting and cost, quality and time planning in large projects helps in adapting plans for emerging or new needs, in line with any transformation in the established budget, quality, resources and costs, to ensure the project is accomplished without compromising defined standards.

Control

It is important to understand that complex and large projects need exceptional control levels, and work is segregated into stages customized as per the unique needs of every project. For instance, the London 2012 Olympics project was segregated into annual phases to offer the clarity and visibility of the focal points necessary for attaining the project's critical, time-related deadlines.[20]

The significance of management in organizations engaged in large projects cannot be overstated. When management of projects is correctly pursued, it aids in running every part of the business smoothly by enabling team members to focus on work that is crucial, without distractions by tasks that move off track and budgets that go out of control.

Apart from this, correctly pursued management empowers the teams involved in large projects to deliver results that actually influence the bottom line of the business, and considerably aids employees in seeing how their work supports its strategic goals. Because large projects consist of varied interrelated tasks that need to be finished within the defined time frame, the importance of saving money and time multiplies because a flaw in the completion of any task affects subsequent tasks and might cause significant delays and budget overruns. Proper planning makes it easy for project managers to ensure that the work is delivered on time and within budget.[22] Effective project management control has the additional major benefits of mitigating the risks of project failure, increasing customer satisfaction, and developing a competitive edge.[23]

An important part of effective control is effective communication.

Communication

Improving internal communication is critical in large projects that have numerous project managers, executives and stakeholders. Working together becomes complex at times but efficient project management techniques and processes make it easy to control the complexity of the collaboration, maximize transparency, and affirm accountability, even when project management professionals are working across departments and teams.[24] Consistent communication with staff, clients, suppliers and investors should be ensured by the project management techniques chosen, along with satisfying the various demands of important stakeholders.

One of the elements that needs to be communicated about is resource management.

Resource management

With the use of suitable project management methodologies, it is easy to map a project's journey from the point of commencement throughout its lifecycle, and to determine in advance where the forecasted expenditure and deadlines are moving. In this way, proper planning and management methods help to optimally allocate available resources, thereby helping to avoid project overspending and delays.[24] Using suitable project management methodologies, clear records of the actual progression of large projects can be kept, resulting in a detailed and deeper understanding of where the resources are spent, which aspects need to be prioritized, and what the timing should be for decisions in case there is a risk of moving off track.

Good management methodologies help in prioritizing business resources, assuring their efficient usage, and defining the schedule, time and budget precisely from the beginning. Thereafter, efficient management of large projects helps in complying with the schedule and managing resources and costs in line with the budget, while also improving productivity and fostering quality work.[25] Whether one is working on a complex project or an easy task, ineffective use of resources can create havoc[26]. By deploying a suitable project management methodology, managers can realize how to efficiently utilize resources to the utmost benefit of a project.

Efficient resource management means that managers can foresee issues before they become critical, prevent bottlenecks and make data-driven decisions.[27] The key benefits of good management methodologies in large complicated projects include improved chances of attaining the desired outcome, a clearer perspective on the project, and better determination of how it adjusts to the central business strategies—and this is especially important in a long-term project.

Long-term project management

It has been found that managing large projects is a daunting job that often goes on for years—since large projects are usually also long-term—during which varied activities need to be performed by optimally using available resources, within a clearly specified time frame while also accommodating market fluctuations that entail changes in client expectations. In addition, under large-scale projects, project members usually face immense pressure to adhere to the project deadlines, budgets, quality measures, and compliance standards and guidelines.

Effective project management is integral to enabling team managers and members to manage, identify, and regulate the quality of key deliverables. Further, only effective management of projects offers the benefits of understanding the various project deliverables and objectives, meeting the demands of prominent stakeholders, efficient deployment of key resources, and the increased probability of attaining the desired outcomes.[28]

HOW ARE LARGE PROJECTS MANAGED IN THE UK?

Failure to consider effective management tactics can lead to project failure and this is evident in the case of Wembley Stadium in the UK. Football Wembley's home had a bumpy start to its life, with conflicts and issues concerning the roof design, the ground processes, and the famous architecture. Apart from this, legal bills mushroomed the cost from what was perceived as a moderate cost overrun in the initial years.[29]

The Scottish Parliament building also presents a good example of ineffective project management that led to failure. In this case, after the commencement of the building work, transformations in the scale, specification and size of the original project plan caused substantial delays and considerable overspending. The original estimated cost of £50m soon jumped to £414 million, making it one of the most unforgettable construction failures ever.[30]

Adequate management of projects that work on a large scale can substantially contribute to reducing costs, and this is evident from the example of the London 2012 Olympics. It was a giant project centered on the renovation of the Stratford area and then developing eight world-class premium facilities, starting with the Olympic Stadium and finishing with the London Aquatics Centre. In this project, success was attained through proper management, not only in the timely completion of the Olympic Stadium but also in its reduced spending — because the project cost was 23.6 percent less than the planned cost.[31]

Apart from the London 2012 Olympics project, the ±£5.5 billion Channel Tunnel project is recognized as one of the most expensive infrastructure projects in the UK since 1985, but suitable management methodology and a hike in the cost of nearly 42.1 percent helped to make the project execution impressively simple. In the construction stage, project managers put more effort into managing the interests and incentives of contractors than into cost containment and, surprisingly, this had a significant positive impact on investors.[32]

From the analysis of these and other projects in the United Kingdom, it is clear that adopting a management methodology that is compatible with the nature of the project, its time scale and needs, as well as market trends and client demands, can ensure effective management of costs and timing without any compromise in the project quality.

Methodologies, definitions, and theories

In order to manage large projects, a wide range of methodologies (often called techniques) are used to aid project management. Joslin and Müller[33] highlighted project success as the crucial factor to measure at the end of a project. The success of a project relies upon several criteria, which encompass time, cost and quality, and include criteria linked with its internal efficiency. In achieving success, project management methodology has gained wider attention in pursuit of enhancing the effectiveness of the project, as well as increasing its success.

Sauser, Reilly and Shenhar[34] explained effective project management methodology using the contingency theory, which emphasizes the relevance of idiosyncratic structures of organization, based on the context. The theory asserts that the accomplishment of a project in its environment is contingent upon its governance through project management methodologies.[34]

Špundak[11] defined *project management methodology* as the set of techniques, procedures, rules and best practices that are used in a project. The guidelines and principles can be applied or tailored as per the situation.

Alternatively, taking the body of knowledge perspective, project management methodology is explained as the set of structured techniques used for solving specific problems. Project management methodologies can be combined to provide a structured way of managing projects, integrating distinct rules and directions with the aim of increasing the probability of successfully delivering the project. Combining project methodologies achieves a higher-quality result, integrates greater control, simplifies the process and improves it.[35]

Since most newer methodologies use classic project management methodologies as a starting point, let us start by looking at those.

Classic project management

Classic project management methodologies are still considered effective. They create an adequate plan for covering all the forthcoming work activities in large projects, determining which tasks need to be performed, and what must be the sequence of application. The plan articulates which tasks must be completed first.

The classic technique also allocates proper resources in accordance with the significance of tasks, and offers proper feedback to and from team members, which not only aids in team building but also helps in timeously monitoring the quality of work finished, and identifying how the deadlines are fulfilled by the team.[15] However, as said before, classic project management techniques must be augmented to handle the current business environment, which is characterized by continuous change, varied market trends and modern technologies.[11]

Agile project management is a very good augmentation or replacement for classic project management.

Agile project management

Agile project management is widely used in the current uncertain and dynamic era of project management. Martin Fowler and sixteen other software developers first used the term *agile* in their *Manifesto for Agile Software Development* in 2001. Its principles overarch many other methodologies. They are deployed in large, modern projects, where they basically break the long run of the project up into shorter *sprints* that facilitate a detailed evaluation of the entire process during the development phase. This in-depth and closer assessment than the classic one promotes adaptive and efficient planning in accordance with the transitions needed in the project as and when a proper shape is attained. All such activities assure robust continual improvement in the developmental phase; they also make the teams more collaborative and organized for generating the best possible results.[5]

Rasnacis and Berzisa[36] claimed agile methodology is the most common methodology for project management processes. The basis of the methodology is that it is collaborative, iterative and data-backed, fast and effective, and values individuals over processes. Instead of the classic planning of the lengthy implementation of the whole project, an agile methodology divides projects into distinct small parts that need to be accomplished by specific dates.

The UK Association for Project Management[37] reported, in regard to accomplishing a project by a specific date, that the London 2012 Olympics construction program faced the issue of the fixed availability of time to accomplish the construction of the infrastructure and venues and, to attain the time requirement, the budget and quality requirements had to be manipulated to deliver the project in less time. Agile project management methodology was used to do this.

There are twenty distinct agile methodologies, wherein the selection and adaptation of the methodology significantly relies upon the distinct type of the project. The methodologies incorporate an adaptive and flexible managerial approach into getting the business results, meeting multiple criteria, and integrating planning according to the business needs.[36]

According to Akhmetshin *et al.*,[38] agile project management methodologies are gaining popularity as flexible approaches. Seventy-one percent of projects claim to be using such flexible approaches due to the diverse benefits associated with them. The crucial benefits are the establishment of a work environment that supports creativity and productivity, quick reaction to the changing requirements of clients, and rapid integration of changes.

Hybrid approaches have gained prominence and are used to enrich the model of a plan-driven process with agile principles, and thus combine the advantages of two or more agile methodologies, some of which are discussed below.[28]

Critical Chain Project Management

The Critical Chain Project Management (CCPM) methodology, based on the theory of constraints,[39] plays a pivotal role in the project planning system. It does so by stating the central tasks and assessing the key stages and final milestones of the project, along with the completion time. A Gantt chart is prominently used in CCPM scheduling and helps keep track of the project activities by facilitating daily comparisons of planned and actual performances.[38] CCPM embraces scheduling, buffer management and planning for multiple projects, and is extensively used to manage large-scale projects because it helps determine the duration of the entire project by working out the longest route or path of the project.[38]

Project Evaluation and Review Technique

The acronym, PERT, refers to Project Evaluation and Review Technique, a software management methodology that is used extensively in several industries and domains. The methodology provides quite complicated, detailed planning scenarios for the project on PERT charts that help the development team clearly visualize the entire process and related outcomes. The key attribute of PERT is that it undertakes an effective and inclusive analysis of the core tasks that are completed within the project. This feature helps the team keep proper track of all the developmental activities and identify weak areas where they need to improvise.

PERT is most suitable for large projects that run over a long term and include a variety of non-routine tasks with transitional stakes.[24]

Critical Path Technique

The Critical Path Technique has attracted attention for managing large projects in the contemporary landscape. It is widely embraced for projects and various tasks that require scheduling and planning of work activities as per the requirements stated in the project brief. It is sometimes used together with the PERT and is considered an excellent means of detecting and confirming the lengthy path for the tasks that need to be performed.[5] The tasks that are likely to happen in a particular trajectory have their distinctive (critically significant) characteristics highlighted so that the tasks can be performed alone instead of as part of a sequential pattern. The critical significance that the technique detects makes it feasible for the development teams in large projects to control the project by completing the vital tasks first. This helps to save time (which is always limited) and to complete a long project with relative ease once the important work is finished.[40]

The Critical Path Technique can be integrated into extremely complicated large projects that encompass a variety of tasks. It is beneficial when the development team does not have a clear idea of which tasks need to be executed at priority so that the project deadline is attained without wasting any important resources and time. This methodology should be considered in domains like software development and construction. Its key advantages include improved scheduling, enhanced planning, and the ability to control bottlenecks. It can be highly supportive in controlling project costs by making sure that project work is completed within the set budgetary parameters. Strict control is essential for successfully completing large projects involving huge budgets, and this technique helps by tying the budget to the sub-tasks instead of to the whole project.[41]

Kanban

First developed by a Toyota engineer, Taiichi Ohno, in the 1940s, kanban is a project management methodology for improving the flow of the assembly process. *Kanban* comes from the Japanese word for *sign* or *visual board.* Since it is based on teamwork and crucial project management principles, it is now recognized as an agile methodology. It can be beneficial for teams looking after large-scale projects because, like the PERT, it renders a visual work plan, but also formulates a sequence of task cards, different from the graphical representation of the PERT.[42]

Kanban is known as an ideal, all-inclusive project management approach to overarching management, as it constantly fosters delivery and ensures less stress for project teams. It emphasizes collaborative work in teams and is developed as a visual project, most beneficial for project plans demanding several steps with numerous teammates. Kanban is appropriate for large projects because it facilitates a visual roadmap that easily shows the different project tasks to be performed with the common motto of project completion and efficacy.[42]

Lean software development

Lean software development methodology focuses on reducing any wastage of resources to maximize customer value. For example, it was used in the London Heathrow Terminal 5 project, in which the methodology generated value and customer satisfaction by optimizing the process flow. It did so by eliminating waste in inventories, products, services, and transportation.[43] The methodology enables organizations to create higher value for customers with fewer resources.

Methodology for PRojects IN Controlled Environments

There is literary evidence that PRINCE2, the acronym for PRojects IN Controlled Environments, is a highly recommended project management methodology, and often used in UK projects.[14] PRINCE2 is a process-based methodology that includes a range of techniques, encompassing project assurance, project support and change control. The Channel Tunnel and London Ambulance projects are key examples of UK projects that used PRINCE2 methodology to overcome the issues of system failures, unorganized projects and huge overspending.[44] The PRINCE2 methodology enables project managers to justify the project by showing that it will create special value, since the planning starts by establishing consumer desires and benefits to support the proper assessment of resources and costs.

Scrum

Scrum was introduced as a framework in 1986 by Takeuchi and Nonaka,[45] and Jeff Sutherland and his team at Easel Corporation adapted it for software development in 1993.[46]

In rugby, a *scrum* is a group of players trying to get the ball. In project management, *scrums* are brief, intense meetings where team members discuss how to remove obstacles, talk about their successes, report how far they have come, decide what the next steps are, and identify any future challenges. The frequent meetings expedite higher-quality product delivery.[47,48] The scrum methodology focuses on building a product with a series of fixed-length iterations, known as *sprints*.

According to Thesing, Feldmann and Burchardt,[49] the scrum methodology is based on prioritization, where objectives are attained in the order of the values assigned to them by the client and, if the requirements change, the values can be reprioritized. The methodology is also applicable when empirical control is needed. The methodology supports teams in producing a software product in sprints of up to 30 days with frequent regular scrum meetings (events), giving agile teams a framework for shipping software in a regular cadence. The sprints generally span one to two weeks and are organized with teams of ten or less.

Scrum methodology breaks the deliverables of the project down into distinct 30-day intervals. It is applicable to collaborative projects with completely dedicated teams that have no heavy constraints of materials, budget or time. However, even if a constraint arises, such as a reduction in budget that often results in shortening the length of the schedule, the scrum technique is still effective. For example, in response to a requirement introduced by the UK government's comprehensive Spending Review 2010, London's Crossrail project managers were asked to reduce the overall budget by £1 billion. In order to overcome this constraint, they used agile project management methodology, mainly the scrum technique, to construct the stations and tunnels sequentially, rather than in parallel, and to extend the date of completion.[37]

Waterfall

The waterfall methodology is a version of exemplary, straightforward planning using the fundamentals of project management.[26] It involves flow planning, based on the principle of breaking the working process into distinct sequential tasks, with specific objectives. The people using the methodology perform the tasks one by one, and do not move further until each task is completed.[38]

The waterfall methodology takes the management concept to a new level. A project is fragmented into sequential phases where each phase relies on the accomplishment of the last one. This management methodology is most suited to straightforward and precise projects where only a few obstacles are anticipated. It can help in planning the entire lifecycle of a large project stepwise and following it systematically. The waterfall technique is the same as agile in terms of linear nature, but is different in being a holistic plan, instead of a set of plan segments of short duration that are revisited for each segment. This technique keeps the planning of large projects simple and permits one to easily exhibit projects. It is not the best choice for more complicated projects or one with several external variables.[26]

Work breakdown structure

Work breakdown structure (WBS) methodology organizes the work into small and easily manageable pieces. According to the Project Management Body of Knowledge (PMBOK),[50] the WBS methodology is hierarchical, deliverable-centered decomposition of the work to be handled by the project teams. The methodology enables project managers who oversee large and complicated projects to divide the scope into smaller segments that can be worked on individually by the team. As the management levels move down in the WBS, more detail and definition are available, rendering a task map for working on a sizable project.[51]

WBS methodology is suitable for both large and small projects because every project involves diverse activities that are interlinked and interdependent.

CONCLUSIONS ABOUT PROJECT MANAGEMENT OF LARGE PROJECTS

Management is central to initiating and executing any project work where the application of planning, controlling and executing processes helps complete the project, produce deliverables, and meet cost, time and quality objectives. However, the current project management landscape requires dynamic or adaptive (agile) management practices that take into account market needs and trends and, overall, the constant changes taking place in the external and internal environment in large projects, which are predominantly lengthy and complex.

A robust decision-making model should be used to select an appropriate procedural model for project management. A decision should be made after weighing aspects like project scope, organizational context, project team characteristics, budget requirements and time constraints. It is most important to understand whether the project scope is small, medium, or large, along with understanding its characteristics, such as the dynamics and uncertainty of the customer requirements. In cases with little uncertainty and a transparent description of the requirements, the classic project management methodology should be used but, in all other cases, an agile methodology is better. An agile methodology helps the project manager to swiftly identify the changes in customer requirements and to react quickly and flexibly to errors.[49]

Chapter summary

This chapter examined the pivotal role of management in various large construction projects in the UK, such as the Channel Tunnel, the Crossrail project, the London 2012 Olympics, the London Heathrow Terminal 5, the Scottish Parliament building and the Wembley Stadium. Robust project management methodologies were deployed in these projects to improve their chances of success, prioritize business resources, and monitor the schedule, time and budget properly to attain the desired outcomes.

I explained how efficient management of such large projects includes managing resources, schedules and costs in line with the budget, along with improving productivity and quality. Distinct project management methodologies have been used to manage large-scale projects in the UK, and agile project management methodologies were prominent among them. An agile methodology helps project managers establish a working environment that supports creativity and productivity, quick reactions to the changing requirements of clients, rapid integration of changes, and better ways of identifying and prioritizing tasks.

I examined project management methodologies like CCPM, lean, PRINCE2, scrum and waterfall in some detail because they are often used to manage large-scale projects in the UK. These methodologies offer structured ways to manage the projects so as to have detailed insights that enable project managers to attain high-quality results, integrate stronger controls, simplify processes, and introduce improvements.

CHAPTER 2

WHY IS EFFECTIVE RISK MANAGEMENT IMPORTANT IN LARGE PROJECTS?

A *project* is characterized by a single specific action that has a logical sequence of activities and controlled and coordinated elements. It is undertaken methodically and progressively, within the constraints of cost, resources and time, to meet defined objectives. Constantly changing external factors have a significant influence on the internal factors of projects, which leads to the emergence of uncertainty and risks.[52] Generally, a project is commenced with an effective strategy but fails in recognizing potential risks that might hamper a project's overall progress. However, an experienced project manager using an effective project management methodology can plan ahead for unexpected threats and risks.[26]

Risk management has therefore become a pivotal element of effective project management, intended to detect and manage the potential and unforeseen risks during the project implementation. *Risk management* is a planned and structured process aimed at helping the project team make appropriate decisions at the right time to identify, classify and quantify the risks so as to manage and control them effectively. The aim of risk management is to ensure optimal project value in terms of time, cost and quality, through balancing the inputs for managing risks and benefiting from this process.[53] Therefore, risk management refers to the process of risk assessment, as well as to the recognition of uncertainty associated with investment-related decisions. In simple terms, risk management is the approach taken by management to reduce or eliminate the impact of risk.[54] Risk management is central to conducting large projects effectively because it helps construction projects reduce a wide range of construction, environmental, financial and other risks.[55] Bahamid and Doh[56] defined risk management as the comprehensive, effective, organized and tailored identification and response to the risk factors that threaten the goals of a project.

Below is a detailed review of the types of risk, the importance of effective risk management in large projects, the iterative risk management process, and risk management tools and techniques.

TYPES OF RISK

A project is prone to diverse risks that can be both internal and external to the project. They are linked to unsuccessful project completion, along with the danger that is beyond the project team's control. *Internal risks* include the risks linked with the internal aspects of the project that are under the control of the team, whereas *external risks* involve the risks that arise outside of the organization and affect the project value to be delivered.[52]

There are different types of risk but, if they are not all managed effectively, the project managers are unlikely to attain the desired objectives of the project by completing it timeously and well. Before understanding the different types of risk and how they can be managed, it is important to understand a little more about risk in general.

A *risk* can be defined as a situation that might cause a deviation from predetermined results. In simple terms, risk can be compared to an injury or fatality that might hamper project activities and prevent the successful completion of the project.

There are risks in every project and risk classification should therefore be accepted as a part of the managerial process for effective risk management and the structured attainment of objectives.[57]

Known and unknown risks

Risks can be classified mainly into two distinct categories: known risks and unknown risks. *Known risks* frequently arise in large projects, causing fluctuations in the project. Due to their nature, the risk source can be identified, and a risk impact and occurrence probability can be assigned (see Figure 1).[58] *Unknown risks* cannot be predicted and, due to their nature, it is quite difficult for project managers to identify them and their potential impact, and to assign a probability to them.[59]

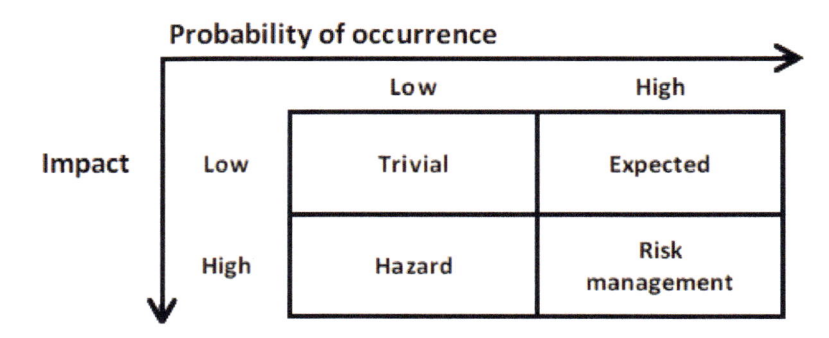

	Probability of occurrence	
Impact	Low	High
Low	Trivial	Expected
High	Hazard	Risk management

Figure 1: Classification of risk (probability and impact)[57]

Fundamental, particular, pure and speculative risks

Hessami[53] broadly classified project risks into fundamental risks, particular risks, pure risks and speculative risks. *Fundamental risks* are external to the project and, if they materialize, they do so on a large scale and cannot be prevented. *Particular risks* are project-specific risks, which are identified within the project parameters and can be controlled effectively during the project implementation; they may be linked with the quality, safety and legal aspects of the project. *Pure risks* have a simple probability of loss or no loss, whereas *speculative risks* have a chance of gain or loss.

Cost, performance and scheduling risks

Project risks can result in various issues linked to cost, performance and scheduling, and may have other adverse consequences for the project. The most common risks are cost, environmental, governance, legal, market, operational, performance, program and strategic risks.[52]

Factor-related risks

Sharma and Gupta[60] categorized the different types of risk in large projects by their related factors; for example, environment-, finance-, management-, resource- and site-related risks, and socio-political risks.

- *Environment-related risks.* Natural disasters and adverse weather conditions are some of the key environmental issues that create obstacles to the successful completion of a project by the project managers.

- *Finance-related risks.* Payment delays, funding problems and a high inflation rate create financial risks for large projects.

- *Management-related issues* in large projects include failure in project planning, inefficient quality planning, lack of experience, poor supervision and site management, a shortage of managers and unclarified job roles.

- *Resource-related factors* involve a wide range of risks, such as adverse project conditions, the breakdown of equipment, delays in receiving materials from suppliers, delays in site possession, inadequate surveys, lack of skilled workers, and low productivity.

- *Site-related risks.* Technical issues in a project arise due to design errors, unclear specifications, and the lack of a proper feasibility study.

- *Socio-political risks.* Permit delays, disputes with labor, acquisition problems, modifications in rules and regulations, and strikes are some of the key risks that arise due to socio-political factors.[60]

Carlos[61] listed twenty-seven factors that lead to project failure. Among them were team mismanagement, overrun costs, scope creep, unrealistic time frames, conflict with stakeholders of the project, lack of effective management, failure to meet user expectations, and wrong decisions.[61] Factor-related risks raised by other authors are discussed below.

Communication-related risks

Project managers often have difficulty in communicating with all the stakeholders of the project. Sound communication is a crucial tool for attaining the desired objectives, and a lack of effective communication adversely affects the quality of the project and has a wide range of repercussions that can have an enormous impact on the project's success.[62]

Culture-related risks

Large projects include people from diverse cultures and the management of different cultures is quite a tedious task for the project manager.[63] Poor management of cultural differences not only arouses conflicts between workers but also diminishes morale, which, in turn, has an adverse impact on the project's success.

Management-related risks

Ineffective management of project risks is the leading cause of project failure. A report published by McKinsey & Company in 2021[64] revealed that most of the issues that cause project failures lead to cost overruns. In this regard, issues like missing the focus of the project, having unclear objectives or lack of business focus, content issues involving the shifting requirements of clients, technical complexities, and skill issues like poor skill sets and unaligned teams are central to the risks of project failure. The risks caused by execution issues pertaining to unrealistic scheduling and reactive planning, and some unexplained causes reportedly lead to cost- and schedule overruns, and benefit shortfalls.[64]

Top management's lack of effective involvement lowers the morale of workers, which affects the work quality and thereby impacts project progress adversely. Lack of commitment by top management is a big problem for the project because it is reflected in difficulty gaining workers' commitment, as well as various other risks, such as difficulty in clearly identifying requirements. If requirements are not clearly identified from the start, the complex nature of a large project means it will probably require huge modifications in its scope. Scope modifications create various risks for project managers in terms of managing the flow of work and maintaining project quality.[65]

Scope-related risks

Frequent changes in project scope are a key risk for construction projects because the implementation of a revised design or scope can have adverse impacts on cost and time resources. This reinforces the need for effectively balancing time, cost and quality trade-offs in projects.[66]

Quality-related risks

A large project faces quality-related risks due to the supply of defective materials by suppliers. This is because, in a large project, it is quite difficult to review the quality of all materials supplied, and the acquisition of defective material creates a barrier to delivering a quality project. Poor site management is one of the biggest problems of large projects because it adversely affects the project's progress as well as its quality.[67]

Finance-related risks

Large projects like those in the construction industry generally face the problem of material wastage, which in turn results in cost overruns. Since large projects typically take quite a long time to complete, this can bring a wide range of financial risks to the project.[67] For instance, inflation can make it difficult to complete the project at the predetermined budget levels. Delays in receiving grants or subsidies from the respective authorities for the project affect the timing of project completion, as well as its financing.

Worker-related risks

Several worker-related factors create risks for the entire project and hamper its quality.[64] Ineffective execution by project managers makes it difficult for the workers to carry out project activities in alignment with the project scope. The absence of project managers from project work can affect the performance of the workers due to a lack of supervision. This might affect the quality of the project. Similarly, lack of trust and cooperation on the project site is a major concern for the project managers. They face huge complexities in managing workers in large projects because the project size requires a large number of employees. The project's size and complexity and the resulting large number of workers create a problem (risk) in effectively defining roles and responsibilities, and this hinders both managers and workers from carrying out project tasks in alignment with the project scope.

All these risks can only be overcome by effective risk management.

WHAT IS EFFECTIVE RISK MANAGEMENT?

Effective risk management is crucial for companies in the modern era to ensure the survival and sustainability of the business in its highly dynamic market conditions.

Foerster and Arnold[68] listed eight steps to ensure an effective risk management program for the benefit of an organization. The first step is concerned with implementing the framework pertaining to risk management, based on the existing risk policies of the company. The second step is to establish the context, i.e., to identify the stakeholders and environment. The third step is concerned with identifying potential and existing risks, along with the related control systems. The fourth step is to evaluate and analyze the identified risks. The fifth step relates to developing risk management strategies. The sixth step is concerned with consultancy and communicating the risk management strategies and other relevant information to everyone in the organization. The seventh step is a detailed review of the risk management strategies, as well as regular monitoring to ensure effectiveness. The last step is to maintain a detailed record of the key documents for future reference.[68]

Effective risk management is important for large projects because they involve massive investments and consume a huge amount of time. Risk management helps project managers enhance project performance by linking project risks and objectives because each risk not only hampers the project activities but also restrains the project managers from attaining the objective of timeously completing a high-quality project. Risk management procedures help project managers identify risks and their impacts on the project's success. Of course, it is also important for the project managers to adopt appropriate risk-mitigating measures to reduce their adverse impact on the project and thus optimize the project's chances of success.[69]

The central purpose of effective risk management in construction projects is linked to providing a safe environment for collaborating with workers and attaining goals. Risk management should introduce a system that assists management to manage the risks in the daily activities, and ensures the activities are effective in a stressless environment, along with enabling the identification of risks by collecting relevant information.[70]

Large projects like construction projects involve a vast range of issues that, due to their complexity and size, might have profound consequences for the project's success. Risk management helps the project managers reduce the risk of these consequences by introducing improvements to the construction process and using resources efficiently.

The project risk management consists of risk management planning, risk identification, qualitative and quantitative risk analysis, risk response planning, risk monitoring and risk control. A wide range of factors affects a large project such as planning, the availability of resources, political factors, and statutory regulations, and thereby enhance the risk that its completion will be unsuccessful. Risk management helps large projects attain project objectives such as environmental sustainability, quality, safety and cost. Following the process of risk identification, assessment, mitigation and control helps project managers improve the project's likelihood of success.[71]

Foerster and Arnold[68] describe construction projects as extremely complex, and linked to uncertainty and risks because of their unique, dynamic, inherently complex and risk-emanative features. These features can have a positive or a negative impact on the multiple stakeholders of the project. Since positive impacts are obviously preferable, effective risk management plays a pivotal role in maintaining the commitment of participants and minimizing adverse impacts on the time, cost and quality objectives of the construction project. Effective risk management helps in controlling the critical elements of a project and keeping track of the threats.

Effective risk management also helps in defining goals better, providing adequate support to the management, detailing the project plan, defining control mechanisms properly, dealing flexibly with uncertainty, enhancing competency[72] and altogether carrying out large projects more efficiently.

Effective risk management helps the management minimize the time taken to complete a large project. Maintaining the time schedule and preventing specification discontent are quite important for large-scale companies around the world.[73] Similarly, the time schedule and budget are quite important for medium-sized enterprises in the market. However, time, scheduling, specification and budget risks can be effectively countered, prevented and mitigated with the help of effective risk management programs to ensure smooth continuity of large-scale projects in the market.[73]

Foerster and Arnold[68] go into detail about the benefits of effective risk management programs for large-scale projects. These take the form of effective strategy planning, establishing a basis for continual improvements in the company, minimizing disruptions, providing a systematic method for managing decision-making, enhancing the knowledge and understanding of management in relation to exposure to risks, boosting profitability via effective job controls, improving cost control, enhancing engagement processes, and reducing litigation by providing contingency plans.[68]

ITERATIVE RISK MANAGEMENT PROCESS

Doval[52] makes it clear that the members of a project team at various levels need to use a structured framework effectively and efficiently in order to identify and manage conscious risks. The use of a structured framework also ensures the planned progress of the project with the least number of surprises or deviations, as per the organization's and the project's objectives. A framework offers an effective and early level of communication of project problems to the stakeholders of the project organization. Its use in the risk management process provides an efficient tool for team building by ensuring the acquisition of and acceptance by the team.[52]

A large project such as a construction project faces various risks during the initial stage of the project. This makes it crucial for the project managers to adopt a sound and strategic risk management approach right from the start for reducing risk and making informed decisions in alignment with the project objectives. Project-related risk analysis at an early stage is also quite pivotal to adopting sound measures for reducing risks and conducting the project successfully. Risk planning in the initial phase of a project can prove fruitful for project success and reduce a wide range of risks.

Petrovic[57] viewed risk management as an iterative process, wherein each risk management facet needs to be planned and followed in every project phase. The iterative process has four key steps: risk identification and analysis, risk assessment, management response and risk monitoring and controlling (see Figure 2).

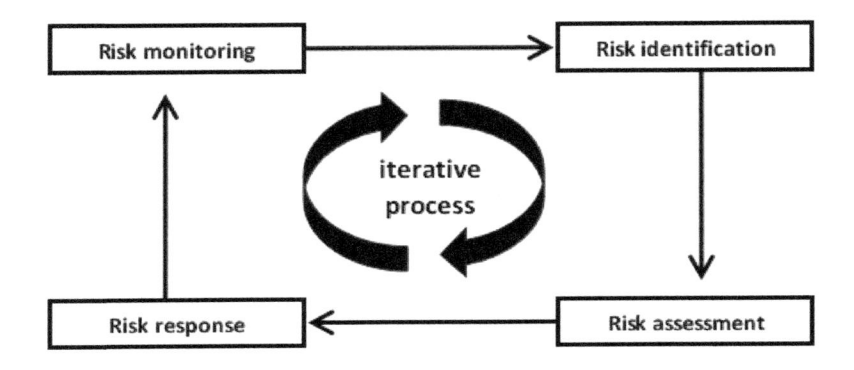

Figure 2: Iterative risk management process[57]

Risk identification

Risk identification is the primary step in the risk management process. The risks identified in previous phases may continue to be identified throughout the project or they may fall away in later phases.

Mhetre, Konnur and Landage[74] suggested various effective risk identification approaches that project managers can adopt, such as brainstorming, interviews with experts, checklists, and analyzing past scenarios. Blasquez et al[75] listed various other techniques that are suitable risk identification in large projects, such as the Delphi method, Hazard and Operability (HAZOP) software, database analysis, and cause and effect diagrams. Szymański[76] presents options for qualitative risk analysis (decision tree analysis: a visual technique that supports decision-making; and model simulations) and quantitative risk analysis (expected monetary value: a statistical technique that helps to compare and evaluate risks; three-point estimate:

applicable to both duration and cost estimates; and sensitivity analysis: also referred to as what-if or simulation analysis).

The main aim of risk identification is to recognize threats that might affect the project's progress and hamper business functions. Although it is impossible for project managers to identify or predict all risks that might occur in the future, reiterating the risk identification step helps them recognize a wide range of risks that can affect the project's success. The project managers can then devise an effective strategy for reducing the impact of these risks so they can successfully complete the project.[77] However, the strategy can only be devised after the risks are assessed.

Risk assessment

The risk assessment is the second step in the risk management process. Its goal is to describe risks and prioritize them according to their impact and the probability of their occurrence.

Applying the probability and impact matrix shown in Figure 3 is one way to do this and to assign an appropriate response. The matrix prompts project managers to discern the probability that any identified risk event might occur, along with its impact on the project.

46 | FAILING PROJECTS IN THE MODERN WORLD

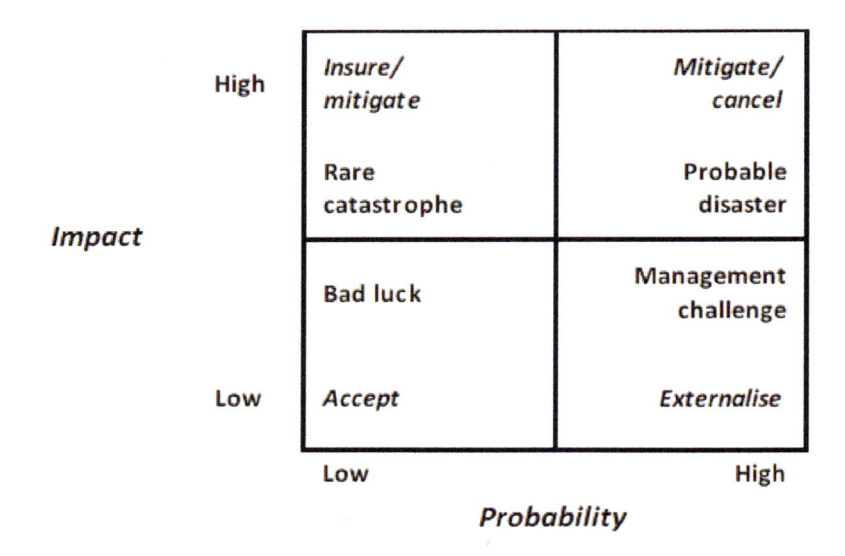

	Insure/ mitigate	Mitigate/ cancel
High	Rare catastrophe	Probable disaster
	Bad luck	Management challenge
Low	Accept	Externalise
	Low	High

Impact (vertical axis) — *Probability* (horizontal axis)

Figure 3: Probability and impact matrix[57]

The risk matrix in Figure 4 is another tool that helps project managers assess the various risks related to the project. It only considers risks that have an adverse impact on the project, but it helps project managers make sound decisions for dealing with them. The project manager can accept (keep) a risk that has little impact on the project or opt for insurance against severe risks.[57]

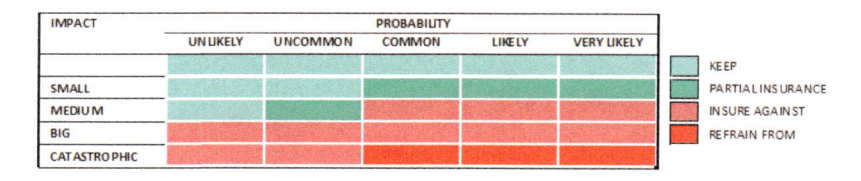

IMPACT	PROBABILITY				
	UNLIKELY	UNCOMMON	COMMON	LIKELY	VERY LIKELY
SMALL					
MEDIUM					
BIG					
CATASTROPHIC					

Legend: KEEP, PARTIAL INSURANCE, INSURE AGAINST, REFRAIN FROM

Figure 4: Risk ratings matrix[57]

Risk assessments can also be conducted using decision trees to help frame problems and evaluate options.[74]

Maintaining a risk register can also be beneficial for risk assessment. The risk register helps in evaluating different types of risk and their impact on the project's success. The risk register should also record various mitigation measures (responses) for effectively addressing the risks in order to attain the project objectives.[57]

Risk response

A risk response is a sound practice or strategy that should be adopted to tackle the risk. There is a wide range of strategies that project managers can adopt for responding to risks. The common response options involve avoidance, transference, mitigation and acceptance. For instance, if the project risk is catastrophic and highly likely, the project manager should avoid the risk by canceling the project because carrying it out might have severe repercussions for the company. The project manager can also transfer project risks that are not quite catastrophic through insurance. If the risk is not severe, the project manager can adopt sound mitigation approaches to reduce it. Risk acceptance involves acknowledging the existence of a low-priority or infrequent risk, but not taking any proactive action to prevent or mitigate it. If the risk occurrence would affect the whole project, the project manager can avoid it by implementing risk mitigation measures like risk monitoring and risk management tools and techniques like keeping an up-to-date risk register.[57]

It is essential to develop an effective risk response plan to address the identified risks, and the plan should include an iterative process, and clearly define the agreed responses.[78]

Risk monitoring

The final step in the iterative risk management process is regular risk monitoring. The continuous monitoring of risks and the strategic measures adopted to overcome them is a significant part of the risk management plan. Failure to constantly review risks affects the risk management outcomes. Risk monitoring not only helps the project manager control a wide range of risks but also assists in detecting and eliminating new risks that arise during project development. It is crucial for the project manager to monitor large project risks that have been identified for the purpose of risk treatment action.[58]

For effective monitoring and control of the risks in large-scale projects, management and the project team should be equipped to make informed decisions about the risks, by iterative evaluation of the effectiveness of the risk response actions, and identification of the risks' characteristics, which can be expected to change in the earlier stages of identification and assessment.[78]

RISK MANAGEMENT TOOLS AND TECHNIQUES

Project managers across the globe have developed effective risk management tools and techniques to ensure their projects are successfully completed within the allocated timeframe and budget.

Project risk management

Project risk management (PRM) as defined by the Association for Project Management (APM)[79] includes management of the tools and techniques that project managers can use to effectively implement risk management practices.

The APM's PRM has four different steps for project managers: identify potential risks, quantify the identified risks, plan for mitigating the risks, and control or monitor the risks. Following these four PRM steps will help project managers effectively protect their projects from exposure to existing and potential risks.[80] PRM is an important risk management tool that employs techniques like root cause analysis to ensure that the project has the desired success without encountering many difficulties.

Root cause analysis

Root cause analysis is a prominent technique in effective risk management that can be used by project managers anywhere in the world in the course of developing large projects where the higher complexities involved lead to problems or risks in the project development. Root cause analysis systematically determines the sources of risks associated with the project. It can be deployed after the problem comes to the surface. Project managers should use the technique to address the key causes of any problem that has arisen, instead of merely identifying the symptoms. This is because root cause analysis helps them adequately counter and prevent the risks by determining the root causes of the occurrence of such problems or issues and then implementing effective risk management responses to meet the project objectives smoothly.[81]

Risk avoidance, transfer, mitigation and acceptance

There are several other techniques that can be used in a project to respond to risks. Banaitiene and Banaitis[71] recommended the four crucial strategies mentioned before for managing risks in construction projects: risk transfer, risk avoidance, risk mitigation or risk acceptance. Risk mitigation and the development of a risk response plan are often the weakest elements of these.

Risk registers

A risk register is used as a risk management tool around the world. In the contemporary market conditions, it enables project managers to conveniently and efficiently monitor and track the existing and potential risks that might hamper the development and success of their projects. The risk register helps the project managers to determine whether the risks, such as delays and over-budget issues, are existing or potential, and to prioritize the risks based on their impact, probability and severity.

The risk register also helps the management identify the owner of each risk, and the nature of the risk, whether potential or existing, thereby providing the basis for making and implementing strategic decisions about mitigation strategies to eliminate or prevent the risk.

Furthermore, the meetings held to compile the risk register effectively share the information with the key stakeholders of the project.[82]

Total quality management

Total quality management or TQM is a well-known technique used by management to manage large projects effectively. TQM helps project managers attain long-term success in the market by meeting the needs of their clients[83] because TQM regularly engages everyone in the organization and they all participate in improving the services, products, culture and processes. Thus, TQM provides security to the project by ensuring the high quality of processes and products to satisfy the customers or clients in the marketplace.

TQM has several elements, such as a customer-focused, integrated system, being process-centered, total employee involvement, continual improvement, effective communication, decision-making on facts, and a systematic or strategic approach. These elements help the project managers to effectively prevent or eliminate the issues of inadequate quality and delays in project delivery.[84]

TQM provides other benefits to the project managers in the form of competitive position, the flexibility to implement change to meet the emerging needs of the market, higher productivity and profitability, the ability to adapt to the regulations imposed by the government, an improved market image, enhanced customer focus, loyalty and satisfaction, the elimination of waste or defects, better cost management, increased customer retention, enhanced job security, improved morale or motivation of the employees, innovative and improved processes, enhanced stakeholder value and satisfaction.[85]

The use of TQM is quite prevalent among large construction projects in order to improve quality assurance, compliance with ISO, and quality control within the sector. The quality level can be assessed and improved by deploying TQM in large construction projects all over the world. For example, a 2011 study conducted on construction contracts in Oman revealed that many companies there were adopting TQM to ensure that the project managers satisfied the needs of the customers adequately and delivered the whole project within the allocated time and budget without compromising its quality.[86]

Just-in-time

The well-known and valuable just-in-time (JIT) technique is deployed by project managers across the globe to effectively monitor large projects and prevent risks that might hamper the success and development of the project in the market. JIT management can align the suppliers of raw materials directly with the schedules related to the production of goods or services.[87] The effective implementation of JIT helps project managers in the construction industry to acquire essential raw materials for the production and development of the whole project at the right time, in the right amounts, thereby preventing the risks of unnecessary wastage and storage of the resources.[88] The inventory system in large and complex projects can be improved with the help of JIT to prevent or eliminate risks concerned with the inflow and outflow of the inventory items.

The JIT technique can be deployed in eight simple, sequential steps: design, manage, pull, establish, fine-tune, build, refine and review. Using these JIT steps in large projects helps in effective risk management because waste can be reduced, obsolete inventory minimized, defective product loss reduced, efficiency improved, productivity enhanced, order changes simplified, production runs minimized, production flow made smoother, and the different types of costs, such as the working capital cost, holding costs, cash investment cost, material spending costs and labor costs, can be trimmed.[89]

Furthermore, JIT is sometimes considered a form of lean production or lean manufacturing because it helps project managers prevent the risks of overproduction, minimize defects in the product, reduce transport costs, minimize waiting times, reduce the requirements for capital, streamline processes, and save resources.

However, it should be noted that JIT has several drawbacks. For instance, it is an expensive and difficult approach to implement successfully.[90]

Strategies

Since there is a negative relationship between project uncertainty and the use of risk management tools or practices in large-scale projects,[91] and since large projects have a wide range of features, it is crucial for project managers to adopt suitable strategic approaches or tools to manage every feature to reduce risk and meet the project objectives satisfactorily. They need to perform scheduled diagnostics and risk assessments. Their risk monitoring strategy should include sound supervision and regular investigation of activities to measure progress. They should also adopt a response planning process, as well as a risk mitigation process, to reduce the project risk effectively.[92]

CONSEQUENCES OF NOT IMPLEMENTING EFFECTIVE RISK MANAGEMENT

If the manager of a large project cannot implement effective risk management, there might be a wide range of consequences, such as missed opportunities and customer dissatisfaction. or the unmanaged risks might adversely affect the delivery and quality of the project.[93,94] Employees might feel demotivated as a result of inadequate risk management and employee turnover might increase. Any delayed delivery or poor quality of the project due to improper risk management will eventually hamper the reputation of the company in the market.[95]

Ineffective risk management might also result in poor governance of the whole project. The management might not be able to adequately control, monitor and supervise the activities or tasks within the project. Another possible consequence of not implementing effective risk management is reckless risk-taking, which in turn degrades the enterprise value in the market.[96] Apart from these consequences, the lack of effective risk management on large projects might lead to insufficient capital, ignorance of the key risks, model errors, agency risks, incremental failure, negative impact on the threat landscape due to sudden shifts or changes in the risks, failure to use a measurement system (leading to mismeasurement of risks), failure to monitor the risks, and lack of communication with top management.[97]

CONCLUSIONS ABOUT THE IMPORTANCE OF RISK MANAGEMENT

From the above analysis, it can be concluded that the risk management of a large project is important for maintaining safety and attaining the objectives of completing the project on time, within budget, and with the desired quality. Effective risk management is therefore beneficial for the companies engaged in large-scale projects to ensure that the project can be delivered without any delay or extra costs as a result of uncertainties or risks.

Due to its nature and complexity, a large project faces a wide range of risks, such as financial risks, failure to define the scope clearly, and ineffective communication. This makes it crucial for the managers of a large project to adopt a sound risk management plan for addressing such risks and attaining the project objectives. The plan should include identifying the various risks and adopting sound tools and techniques to achieve risk reduction targets. Project managers can choose to accept, avoid, mitigate or transfer risks.

The choice of risk management practices, tools and techniques is important for ensuring the effectiveness of the risk management. Tools like risk registers, and techniques like TQM, JIT, PRM and root cause analysis are crucial across the globe for preventing, identifying and mitigating the existing and potential risks associated with large projects.

CHAPTER 3

WHY HAVE CONSTRUCTION PROJECTS FAILED IN THE UNITED KINGDOM?

The construction industry in the UK plays a significant role in the country's economy and is a major contributor to its GDP. According to a study conducted by Urbański, Haque and Oino,[98] the construction sector contributes around £90 billion to the UK economy annually. This means that, if a construction project fails, the economy of the UK is impacted.[98] Hence, there is a need to reduce the number of project failures, starting with finding out why they failed.

WHY I INVESTIGATED THE PROBLEMS

Many factors affect the operations of the industry, such as changing government policies, upgrades in technology, and the changing needs and demands of the laborers. Finding out which factors really caused construction projects to fail seemed to me to be one of the most important and interesting topics that needed to be researched in order to improve the industry I work in. Another reason was that I wanted to enhance my knowledge and understanding of how the construction industry works, and especially of its technology-, management-, and workforce-related issues.

WHAT WERE THE PROBLEMS?

Overbudgeting, lack of available skilled workers, and lack of guidance from the leaders and managers were among the identified causes of project failure.[99] Nguyen and Watanabe[63] believed projects in the construction sector were failing due to lack of communication, poor management, insufficient availability of resources, and lack of technology. Many construction projects, such as the Channel Tunnel, were over budget for these reasons.

Although the main reasons behind the failure of the projects were lack of communication, inappropriate tracking and monitoring, and poor management of project work, projects in the conduction industry faced other problems that were unique to the situation of that particular project, and project managers often could not respond to them effectively and timeously because they could not identify the problems.

With advancements in technology, there were effective solutions to the majority of the problems, and the most effective solutions were the use of artificial intelligence (AI) technology and building information modeling.[63] But problems were still taking place.

Without first-hand information, many of the main causes were not identified.

HOW OTHER PEOPLE SAW THE PROBLEMS

Before I could go any further, I needed to know what other scholars and researchers thought about the problems in the UK construction industry so I reviewed the literature they had written about the project management concepts, theories, and risks (or challenges); the contemporary scenario of failed projects in the construction industry; the vital factors causing projects to be unsuccessful; and possible measures, and their efficacy, to eliminate factors leading to failure.

Project managers have many different ways of conceptualizing what they do.

Project management concepts

According to the APM,[37] project management is the application or deployment of crucial procedures, skills, and methodological solutions, as well as a knowledge base for attaining particular objectives associated with the undertaken project in accordance with the criteria for project acceptance within all the parameters agreed by the stakeholders.

Koskela and Howell[12] described project management as a set of procedures comprising initiation, plan development, and execution, as well as proper controlling and closing procedures. They emphasized that project management is performed by the project team with several goals: The primary goal is to produce the target product; secondary goals involve minimization of cost and enhancement of utilization level; and the final goals of the project are to satisfy customer needs related to dependability, quality, and flexibility.[12]

In the opinion of Teslia *et al.*,[100] project management refers to planning the project, and handling and arranging the project team within the organization with the primary motive of directing and helping the team members to implement the project effectively and accomplish its goals.

Walker[14] saw six stages in the project management lifecycle: definition, initiation, planning, execution, monitoring and control, and closure. The first stage refers to the *definition* of the project, with the focus on identifying the objectives and prioritizing the work for the success of the project. The second stage refers to project *initiation*, in which all the factors that may affect the project are considered, and all the elements needed to build up the project are arranged. The third stage refers to the project *planning*, where detailed instructions and a blueprint of the project are prepared, including resources and prices. The fourth stage focuses on project *execution*; that is, work takes place to deliver the project on time. The fifth stage refers to project *monitoring and control*, where the focus is on tracking the work in process to ensure that everything is taking place in the right direction. The last stage is project *closure*, where steps are taken for the client to accept the deliverables and to disband all the factors that were required to run the project.[14]

Project management risks

Rabechini Junior and Monteiro de Carvalho[101] mentioned several kinds of risk that tend to be associated with the projects that an organization undertakes. The crucial risks linked to project management are financial risks, resource scarcities, schedule-management-related risks, inadequate competencies, and deficient ability to manage the project.[101]

Several risks are described below, with the theories and measures that can be used to counter them.

Communication-related risks

Sauser, Reilly and Shenhar[34] expressed concern about the communication-related risks that tend to arise during the project management process.

Contingency theory

Sauser *et al.*[34] argued that the contingency theory is important in understanding the risks, associated effects, and reasons for a project's failure. Their study demonstrated that applying this theory helps project managers investigate the level of fit or misfit among the characteristics of the project, the method undertaken, and the managerial approach.

Vidal *et al.*[102] explained how the application of the contingency theory in business operations helps managers and leaders determine the effectiveness of their strategy in the different situations of the business project. This theory helps in the effective management of the project work in order to maximize the skills and abilities of the team members so as to complete the project on time.[102]

Diffusion theory

Kreps[103] suggested that implementing the diffusion theory is the best way to ensure that advanced technologies, ideologies and thoughts are diffused or exchanged during communication between the team members in order to make communication productive for successful project management.

Wani and Ali[104] asserted that the diffusion theory plays a significant role in this changing world, where the focus of every business is to grow by adopting and implementing innovative ideas into its operations in order to accomplish its goals and objectives. Applying the diffusion theory in the process of project management helps produce new ideas or methods to deal with obstacles in the most effective way, and to complete the project on time.[104]

Resource-based risks

Sauser *et al.*[34] pointed out the challenge of resource scarcities or how inadequate management of resources hampers the success of the project management.

To mitigate this challenge, Almarri and Gardiner[105] proposed a resource-based theory that emphasizes the acquisition of sufficient resource capability to fulfill the requirements for successfully managing the project and enjoying competitive benefits.[105]

Scheduling risks

Taylor[106] emphasized the scheduling-oriented challenges confronted by the project team since they can lead to a project's failure to be accomplished promptly.

To mitigate this challenge, Aliyu[107] asserted that the critical path method and Gantt chart techniques support the optimum scheduling of the critical activities linked to the project.

Queuing theory

Rashid *et al.*[108] preferred to apply the queuing theory, a mathematical approach that covers a wide range of businesses because it helps in designing data networks and in providing fast customer services by improving warehouse shipments. The queuing theory focuses on components such as the arrival process and the service process to perform all the operations effectively and efficiently.[108]

Haghighinejad *et al.*[109] noted that the application of queuing theory helps a business make appropriate decisions related to ongoing projects and helps in building a cost-effective workflow system. It is an efficient operational management tool that contributes to solving problems related to staffing, scheduling, and customer service. The primary aim of this theory is to work so as to achieve an efficient and affordable balance.[109]

Human resource risks

Human resource management tends to be a serious challenge for project managers in the present era where there are risks associated with teamwork, staff recruitment, working in virtual teams, labor-related issues, cultural challenges, and cross-cultural leadership skills. The key factors affecting the efficiency and performance of project managers in the majority of projects are a lack of financial incentive strategies, the lack of a clear relationship between performance and standards, the lack of a means to align intangible outcomes, the existence of a time-consuming decision-making process by the clients, systematic bias, the lack of a remuneration scale, delays in responding to requests for information, and the impact of culture on project outcomes.[110]

Cultural challenges

The cultural background of the team members and the environment they work in have critical effects on the performance of the teams. Consequently, it is essential that the project managers consider the values and culture of the individual team members along with the key aspects of the working conditions relating to the project.[110]

An important challenge in this regard is becoming an employer of preference. This makes it essential for the project managers to help the organization create an environment (an organizational culture) that is appreciated by the employees so that they choose to work there for the long term. This challenge highlights the extent to which human capital management is conducted by both the human resource department and the project manager, and affects the best practices for the organization, performance, and efficiency.[110]

Technological risks

Tabassi *et al.*[110] saw technological innovation as a challenge for the human resource managers in project management who must facilitate the training in the use of alternative technologies that will optimize the overall success of the project.[110]

Project management theories

In today's changeable world, several theories about methods of managing projects to avoid risks and challenges have been introduced. According to Shankarmani *et al.*,[111] agile methodologies not only welcome any changes to the project during its development, but also ensure fast project management with real-time updates, quality maintenance and low-cost investments.

Scrum methodology

As per Tytkowska, Werner and Bach,[112] scrum technology used in the management of projects validates the flexibility of time and the budget that, in turn, ensures tasks are controlled effectively by project owners to manage the project work's efficacy.

Supervision

Tabassi *et al.*[110] stressed that project managers tend to encounter several challenges in everyday activities that require effective supervision to reduce rejection, flawed outcomes, and rework activities. Additional challenges include shortage of materials, lack of a skilled labor force, lack of clarity and completeness in technical specifications, frequent changes in the orders during execution, and difficulty establishing key performance indicators for evaluating suppliers.[110]

Scope management

It is essential for project managers to implement effective scope management in the early stages of the project management plan so that the risks associated with work interruption caused by an unexpected scope creep can be addressed.[104]

Contemporary construction project failures

According to Gamil and Rahman,[113] the construction industry has always been fragmented, complicated, and dynamic in the market environment, which is why project management in this industry remains highly challenging. Their study of systematic review as the methodological approach revealed that there is an increasing incidence of project failure in the construction industry due to poor management of the communication process, accidents, faults, budget overruns, schedule overruns, and improper planning and coordination.[113] However, the Gamil and Rahman[113] study provided only generalized insights into the project failures in the construction industry.

More usefully, an article by Thomas Owen of Re-Space[114] gave quantitative or statistical information about the UK construction industry's unsuccessful projects. For example, construction projects of almost all categories are unsuccessful in adhering to their budget and schedule: Up to 39% of rail projects and up to 61% of aerospace projects could not be completed on time within the set budget.[114] Notably, the HS2 (High Speed 2) Crossrail infrastructure project was due to be opened in 2018 but that still had not happened by the end of 2023. The overall budget in 2009 was £37.5 billion but, by 2021, the project cost had doubled to £72 billion.[114]

Project leadership

As Shenhar[115] said, the UK construction industry is the major contributor to the economy of the UK, but faces problems in the work process that lead to the failure of construction projects, mainly due to a lack of project leadership. This is because the construction team should be given direction by an experienced project leader who has sound decision-making skills: Without supervision by managers and leaders, issues arise such as the improper use of raw materials and resources, and not being able to accomplish daily targets, all of which contribute to failure.[115]

Budgeting

A 2019 article in *BBC News*[116] mentioned that, due to the construction business's collapse, major construction projects like Newcastle University's £39m extension and £25m sports center were not successful.[116] According to the *Construction Industry News*,[117] the Lincoln Eastern Bypass has also not yet been successfully constructed due to budget overruns and environmental factors like floods. Other construction projects, like the Holyrood fiasco and the Millennium Dome, have also been unsuccessful because they overran the budget.

A 2020 KPMG survey[117] showed that 69% of projects in the construction industry tend to be over-budgeted by more than 10%. A project that is considered unsuccessful (because it is one of the most over-budget construction projects in the UK) is the Channel Tunnel project that was constructed to connect England to France. The project completion was delayed by a year. The reason for the delay and the project's failure was poor communication between the two sides. It is one of the most expensive projects in history.[117]

Performance management and control

Elliott[118] highlighted that effective performance management and control are major factors that lead to improvements in project management and contribute to reducing the cost of the project and delivering more effective outcomes.[119]

Artificial intelligence

Liao and Ai Lin Teo[120] believed that the use of AI by the construction and engineering project management sector is most helpful for improving working techniques, and makes their results more effective. Projects that use AI technology have shown significant improvement in meeting quality- and time-related objectives.[120]

Collaboration

Chen *et al.*[121] explained that construction projects are highly complicated and require cooperative work by the multiple project stakeholders to attain unified goals and vision. The common stakeholders associated with construction projects include designers, project owners, planners, engineers from various disciplines, material suppliers, contractors and planners. With the increases in the complexity and scale of the construction projects, the number of stakeholders and the extent of the complexity in the stakeholder relationships involved in partnerships has also tended to increase. Nevertheless, successful outcomes can be attained from collaboration, effectively capitalizing on the abilities, expert knowledge, skills and experience of the members of the various teams and organizations associated with the project. However, project managers' understanding of the various aspects of partnership relationships is often inadequate. This lack of clear understanding usually leads to the failure of the stakeholders to work collaboratively for the success of the project.[121]

Clear understanding is achieved by clear communication.

Communication

Effective and open communication is a critical concern for the success of a construction project. *Open communication* refers to an open and transparent exchange of resources like skills, knowledge, ideas and technology through a range of effective tools and communication approaches. Consistent, two-way communication facilitates a free flow of innovation, thereby serving as an effective avenue for sharing information about new processes and high-level information that is not available outside the collaborative arrangements of the project. Several studies on construction partnerships emphasize the ways in which effective communication in a partnership is deemed essential for the success of the partnership. Conversely, inadequate communication leads to the failure of construction partnerships.[122]

Partnerships that aim only at conveying information between the site office and head office tend to succumb to failure. Problems at the project site should be addressed without delay and the outcomes should be communicated across all levels of the workforce hierarchy. Without open and honest communication by the project partners, there is not enough resilience in the partnership for the problems to be managed effectively. The project partners also tend to struggle to integrate their respective resources to accomplish the common goals of the project.[123] Poor integration places the project partners in serious and problematic positions because it leads to inefficient implementation of the project, and might also lead to dissolution of the partnership and commercial uncertainty for the parties to the project.

The primary stakeholders in various construction partnerships are the owners and construction contractors, and all these stakeholders tend to have their own cultural values and beliefs that demand cooperation and harmony for effective coordination of resources, time and communication.[121]

To prevent failure in partnerships, it is essential that there is continuous exchange of information about the project between the partners. It is also essential for the communication channels to reflect the organizational structures and specify the information that flows through each channel from the top down and the bottom up. The frequency and timing of communication must help the decision makers receive timely information but not overload them with unnecessary information. Additionally, there must be a transparent flow of communication from the management across the entire organization to ensure that the workers feel respected and engaged.[123]

Transparent communication helps avoid conflict and unresolved arguments.

Conflict management

A major factor identified by Chen *et al.*[121] as the cause of failed projects in the construction industry is unresolved arguments, which, if not addressed at the right time, create a serious negative impact on the operations and overall performance of the project.

The project partners are especially concerned about the successful completion of their project, i.e., that the project delivers on its agreed specifications and quality, and meets key budget and scheduling limitations. The successful completion of a project largely depends on the effectiveness with which the partner organizations address conflicts and maintain reliable, effective, and long-term professional working relationships. Construction projects tend to have challenging working conditions because of the numerous complex operations that are subjected to fluctuations caused by a diverse workforce, challenging working conditions and a serious risk of occupational injury. If such challenges are not addressed quickly, even minor conflicts might aggravate them and have a serious negative impact on the outcomes of the project.[124]

Factors that make projects unsuccessful

Soliman[125] used a questionnaire to interrogate client supervision members and contractors about the problems in construction projects. The findings reflected that the poor quality of documents, inadequate progress in construction site meetings, use of outdated systems, and team leaders with a low level of work experience caused the poor management of a construction project.[125] However, this study did not reflect the broader factors that lead to construction projects' failure.

These were reflected in a study by Gamil and Rahman,[113] which included a systematic literature review and revealed that the success of construction projects is affected by inadequate communication, technological malfunctioning, inaccessible information, unclear project objectives, poor coordination, improper designing or drawing, and contractual barriers.[113] However, this study also did not disclose the major reasons for the failure of the UK construction industry's projects.

Nagarkatte and Oley[39] revealed that the theory of constraints helps determine the element that places the highest limitation on attaining the project goal based on its full potential. They advised project managers to enhance that limiting element or constraint gradually to make it no longer a constraint.[39]

Budget overruns

An article by Edgington[126] expressed the opinion that the most common and critical factor that causes construction projects in the UK to fail is cost or budget overruns. For example, the HS2, a controversial rail project designed to connect Manchester, Birmingham, London and Leeds, was confronted by double the cost budgeted for, up to £106 billion. The ground conditions were underestimated because not all the necessary surveys were done properly in Phase 2: They had to purchase land and properties along the route, underground piping under London had to be rerouted, and roads and even rivers had to be removed. All this added to the cost and also meant the project could not be completed on time.

Resource mismanagement

Factors like disputes, confusion, overbudgeting, delays, and skill shortage cause projects to be delayed. However, by steadily proceeding with their limited resources, extreme effort and the best use of resources, some projects can still be accomplished. The Scottish Parliament building is an example of a project that was postponed because of cost overruns but then the gradual introduction of new ways of managing resources made it possible to complete it late.[127]

Carlos[61] asserted that insufficient resources and lack of communication are the major factors that negatively affect project operations, whereas work performed in coordination and support of each other provides effective results. Effective planning should be the first step in the project process because it solves most of the problems or difficulties in the ongoing project process. Poor day-to-day management is another reason for the failure of the project because effective management of the daily completion of targets, along with proper monitoring and tracking, contributes to achieving the planned outcomes.[61]

Scheduling mismanagement

Another project, the construction of a channel tunnel for traveling to Brussels and Paris from London was also unsuccessful in meeting its planned delivery time because of factors such as schedule mismanagement and cost overruns up to £2 billion.[126] Edgington's[126] findings suggest that resource constraints and their poor management led to the project failure

Technological malfunctions

It is critical that technological organizations establish clear priorities at the portfolio level. At any time, the composition of a technology portfolio can change, based on new project initiation, technical or commercial failures, competing priorities or financial concerns. Effective technology portfolio management is achieved by establishing clear, well-aligned performance metrics relevant to the business and business model. An effective project manager provides the tools and disciplines to accurately track these metrics and enable objective decision-making by senior management.

The next section looks at what measures have been implemented to address the factors that lead to project failure and how well they worked.

Possible measures to eliminate the causes of project failure

Because the construction industry is such an important part of the UK economy, the UK government introduced construction management strategies in 2011.[128]

Building information modeling

According to McPartland,[128] one of the crucial measures proposed by the UK government for construction businesses to manage projects successfully was building information modeling (BIM). Through BIM, it is possible for the project team to collaborate effectively and to use digital technology to raise efficacy and productivity. With the BIM solution, data-driven decision-making has also become possible, and offers the benefits of offsetting carbon impacts and costing the overall construction project.[128]

Rodgers, Chong and Preece[129] stated that BIM is an effective digital transformation approach used in the architecture, engineering, and construction sector. BIM is helpful in the planning phase of a construction project, which is the first and foremost step, which can be most effectively and efficiently conducted with the help of a 3-D design of construction models.[129]

Procurement models

In 2014, the UK government[130] disclosed a new construction strategy to streamline procedures and save costs up to an estimated 20% in construction procurement. The government sought to incorporate innovative procurement models, such as cost-led procurement, integrated project insurance and two-stage open-book procurement. Such models can minimize construction project costs and reduce overall risk through enhanced supply chain relations and innovation.[130]

Research and development

In response to a 2019 UK government policy paper,[131] with the enhancement of digital infrastructure through public investment of up to £1 billion and emerging innovative construction technologies, the UK construction industry put significant effort into partnership with government to improve the management of construction projects. The industry invested approximately £406 million in technical education, science learning and engineering skill development, and adopted more significant strategies to eliminate the major factors of project failure.

Anderson[132] provided compelling evidence that the adoption of such measures was essential for upgrading the country's project management scenario, although he had reservations about the considerable challenges still confronting the UK construction industry, in order to please all the stakeholders. Among these challenges were proper funding of collaborative projects, value addition for all taxpayers, the use of minimal energy, performance management of constructed buildings, and better job opportunities. All these challenges require management.[132]

What excited Anderson,[132] however, was a promise of an £18 million investment by the Research and Innovation department of the UK to revolutionize the construction industry with new and advanced technologies. This guarantee enabled four research projects to be undertaken to enhance quality and cost-effectiveness. The integration of building elements manufacturing with the use of digital hybrid concrete printing technology contributes to effective management of construction projects. The application of on-site and off-site multi-robot automated building construction technology has also proved significant in effectively managing construction projects.[132]

Another successful development was the increased use of AI.

Artificial intelligence.

The construction industry used AI to ensure the effective movement of goods and services. The use of AI technology in the operations of construction projects also contributes to reducing the workload, and ensures the work performed with the use of this technology is the most appropriate, making the chances of mistakes much lower.[131]

Computer-aided design.

Lorek[133] explained that, in the past, blueprints were designed and drawn by hand. Then came computer-aided design (CAD), which helped in designing a 2-D planning model. Now, it has been found that plans designed with the help of a 3-D model promote a better explanation and understanding of the project work. The chances of mistakes are reduced, and a 3-D model facilitates the effective and timely completion of the project.[133]

The theory of inventive problem-solving

Delgado-Hernandez and Aspinwall[134] recommended the theory of inventive problem-solving (TRIZ) because it supports multiple tools for resolving situational complexities and problems. These tools include a survey of a focus group of customers to determine their needs for projects to be accomplished successfully.

A matrix diagram, brainstorming, critical path methodology, PERT and CAD designing software are essential tools and measures for construction projects to be completed without delay by managing resources and project scope effectively.[134,135]

Workforce-related measures

In the opinion of Damoah and Kumi,[136] the construction industry could reduce delays and failure in the projects by focusing on the age of its workforce and skills shortage factors. The focus should be on hiring skilled employees who do not belong to the age group of 50 and above because older employees lack speed and quality. The industry should concentrate on improving the skills and abilities of young employees who can work quickly and maintain quality.[136]

The UK Health and Safety Executive construction statistics from 2019[137] suggest the high numbers of accidents recorded in construction work indicate the construction industry should also take effective measures to improve the health and safety of its workforce.

Waste reduction measures

From the viewpoint of Halliday,[138] reusing material instead of dumping it as waste is an essential and cost-effective approach for managing the work on construction sites, because it helps in reducing the risk of overbudgeting. Reusing material enables tasks to be completed on time, if it involves the use of material with little or no change in its actual format. Construction demolition materials can be used for land reclamation and road surfacing. Coal fly ash is not a waste material; rather, it is valuable for replacing the percentage of cement in concrete mix and rendering materials.[138]

Conclusions from the literature review

Overall, my literature review reflected a lack of first-hand information, empirical insight, and statistical examination of the predictor factors for the failure of construction projects in the UK and relevant measures to mitigate them.

Conclusions from my research

Since a scarcity of first-hand information made it difficult for project managers to decide which measures to employ, I decided to undertake research in the form of a survey and statistical tests to address the knowledge gap.

The variables I chose for my quantitative regression analysis (see Appendix A, How I investigated the problems in the UK construction industry) were *Influence on construction project failure, Scheduling-oriented challenges, Resource scarcities,* and *Technological malfunctions.*

Although the results of the statistical tests showed that, of all the three independent variables, only resource scarcities significantly influence project failures in the UK construction sector, all three of these problems are discussed below.

Resource scarcity

Resources can be anything from materials to equipment to staff power. A lack of them in project management need not spell disaster. In fact, a key part of a project manager's role is finding a way to make things work with the resources they have available and without exceeding the budget.

Good project managers can navigate through their projects and supply their deliverables on time. They know how to make the most of the material and equipment resources available and get more out of their staff.

There are many reasons why resources could be stretched and it is the job of project managers and their teams to figure out the next move.

What causes a lack of resources in project management?

The saying "everything that *can* go wrong *will* go wrong" is never truer than when you undertake a new project. If only everything could operate as planned all the time, we would never have any issues. Unfortunately, that is not the reality that project managers face. Problems rear their ugly heads as often as they can if the manager is not prepared for them. Prevention is the best way to deal with a lack of resources.

How can a lack of resources be prevented?

If you can spot potential issues early enough, you can work on a plan to prevent the worst from happening. Of course, it is not always going to be possible, but forestalling a lack of resources in project management is a big part of the job.

The easiest way to do this is by developing a method of monitoring the resources while the project is under way. At the start of the project, you should compile documents on resource allocation, budget limitations, and the project schedules. It may be worth splitting the project into phases that are easier to monitor and make it easier to control the risks.

Here are some things you should consider doing at the start of the project:

- communicate the project plan and plan the project communication protocols
- create a RAID log (risks, actions, issues, decisions)
- define the scope of the project

- develop a baseline budget
- estimate activity levels
- forecast and make estimations based on previous projects
- identify any project lags
- set deadlines
- understand the deliverables
- use project management software.

Companies are slowly moving towards increased cooperation using technology and integrated schemes, working on shorter timelines. Thus, the Lean-Agile methodology has emerged in recent times as one of the key management frameworks. Great project managers should use it to balance the requirements and demands of all of their team members, keep their team on track, and effectively communicate with their workers and clients. It is a complicated effort, and so it is not surprising that most project managers run into scheduling problems along the road.

Scheduling-oriented challenges

Common problems in project scheduling are budgetary issues, miscommunication, no mechanism for feedback, scope creep, sub-optimal resource allocation, unclear goals, and unrealistic deadlines and planning.

One of the most significant dangers to any project is unclear or inefficient communication.

Communication issues in project scheduling

We are often faced with events in our lives that result in miscommunication or in someone's misconstruing something we said to them. Generally, these events result in conflict and a loss of trust, and hamper cooperation. Inappropriate and improper delegation on the part of the project manager may lead to confusion on the part of the stakeholders. In the project management sphere, it is the product that is affected.

Unclear goals, and unrealistic deadlines and planning

The absence of explicit project goals and a mismatch of goals with essential business objectives are key challenges in project scheduling, which frequently emerge as a result of inadequate planning. This is not always the project manager's fault. Organizations often fail to devote enough time and effort to planning projects effectively and ensuring that those initiatives are linked with a corporate strategy or roadmap. Project managers must then act as a bridge for communication between their team members and the organization's executives. A project manager must also devise a system for measuring project progress by establishing project milestones and quality testing. A defined set of objectives will assist a team to grow and help project managers justify their vision to stakeholders.

Technological malfunctions

Organizations that develop complex technologies face unique challenges that project management can help solve. Even smaller technological companies, including startups, are beginning to make use of project management, while bigger, technology-centric organizations are creating formal, fully staffed project management offices (PMOs). However, some of these technology companies' investments into project management and PMOs do not deliver the intended value. I believe the main reason for this failure is senior management's indifference.

While one of the duties of a good project manager is to influence skeptical stakeholders, no amount of project management skill will overcome senior management's indifference. Moreover, while there should be a healthy tension between project management and functions such as marketing, engineering, finance and operations, project managers need the right amount of organizational visibility, influence and authority in decision-making. Good project managers attain this when they have the full support and commitment of senior management.

Technological companies also fall into the trap of believing that their technology is so great that it will overcome challenges in the schedule, compliance, cost, performance and scope. Construction companies that use technology in their projects can also fall into this trap.

Two of the most insidious problems that can rapidly cripple technology-driven projects are poorly defined scope before the launch and ineffective scope management during execution. A good project manager will properly initiate any project by understanding the bigger business picture and the technical challenges and issues. He or she will also take the time to identify and quantify technical, operational and commercial project risks and incorporate risk management into planning.

The next chapter goes into detail about what good project managers do.

CHAPTER 4

WHAT SHOULD PROJECT MANAGERS DO?

It is becoming ever clearer that the landscape of project management is changing with every passing minute. It is of paramount importance to keep pace with new trends if you are interested in understanding the benefits of project management as it brings direction and great team leadership to projects.

This chapter focuses on the elements that enable project managers to be great team leaders and to direct projects in order to attain project objectives. The chapter highlights the importance of teamwork and project managers' competence to manage the project team, the principles of teamwork, project managers' communication and information skills, technical or professional competencies, and situational management and leadership, as well as virtual course completion and evaluation by teams.

WHAT IS PROJECT TEAM MANAGEMENT?

Project team management refers to the ability of project managers to ensure that people coordinate to carry out their tasks and optimally access resources to attain the objectives set for completing the project successfully.[139] It involves good communication, working closely with other managers, a positive culture, and clear objective setting to stay organized and efficient.

Why is teamwork important?

Teamwork enhances team members' morale, which improves their efficiency and enables them to do work of better quality than each can achieve alone. Team brainstorming leads to innovative ideas. Teamwork improves working relationships and creates healthy competition.[140] It also improves efficiency by allowing complex tasks to be split into simpler forms, each performed by a part of the team.[141] Therefore teamwork plays a crucial role in the success of a project.

Teamwork is central to the execution of projects, increasing the productivity and efficiency of project phases and activities while positively contributing to project success and timely completion — and project managers play a major role in selecting and managing the team. Strategic alignment is also a crucial part of project team management because the team efforts and intentions of team members need to be in line with project outcomes and long-term organizational objectives. The strategic alignment should include stakeholders from the start so that bureaucratic pushback can be minimized, the scope of the project can be clarified, and ambiguities can be reduced or eliminated in the execution phase, which in turn reduces the probability of project failure.[142] The teamwork developed while working on different projects leads to timely completion of projects, high satisfaction among team members and a healthy or productive work environment as a result of increased harmony, understanding and coordination.

Project team members need to fulfill varied responsibilities to provide expertise, establish the overall business requirements and the specific project objectives, work with fellow team members to meet the individual and team deliverables that contribute to attaining the project objectives, and document the procedure.[143]

Teamwork principles

The main principles of teamwork include accountability, clarity, commitment, conflict resolution, creativity and trust.[144] Considering teamwork principles helps project managers reduce the risk of project failure because it facilitates meeting targets for the timeous delivery of the project by defining the scope to every team member, using resources optimally, and enabling team members to work in alignment with the project scope and objectives.[145]

Team management weaknesses

Some of the critical factors that contribute to project failure are team weaknesses like ineffective definition and fulfillment of roles and responsibilities, lack of proper communication with team members, lack of proper support to team members, and failure to define scope to the team members.

Teams and working groups

A *team* is a group of associated people[146] who work together to achieve a common goal.[147] A *working group* refers to two or more persons with different accomplishments who work independently to achieve their common goal.[147] A team has particular objectives for completing a specific task, but a working group works for the organizational objective. Team members are encouraged to assume their assigned, diversified roles, whereas working group roles are decided based on their professional qualifications. In a team, all the members are mutually accountable, whereas, in group work, individuals are held accountable.[148]

Project managers can encourage good teamwork by creating an open and transparent work environment, delegating and scheduling tasks, and exhibiting strong leadership skills by focusing on supporting team members and constantly motivating them. It requires team building.[142]

Team building

Tuckman identified five phases of team building: forming, storming, norming, performing and adjourning, which are explained in Table 1:

Table 1: Team building phases [149,150]

Phase	Explanation
Forming	This is a significant stage for initial team development, in which the focus is to develop good relationships and define the mission together. Responsibilities are set and goals are discussed.
Storming	In this stage, conflict situations arise due to cultural differences, opposing opinions, competition, challenging authority, and arguments.
Norming	In the norming stage, the relationships of team members improve because they establish rules, start communicating effectively, and start developing trust and team spirit.
Performing	This stage improves trust among team members and improves their ability to make decisions on their own, without any supervision by others. This helps them resolve disputes and boosts their confidence that they can achieve the team goals.
Adjourning	This is the last stage, for breaking down the team after completing the project.

The project leader or manager plays a pivotal role in all five of these stages by ensuring proper team formation and management. In the *forming* stage, the project manager's part in team actions includes making it possible for members to learn to know and interact with each other, providing essential information, emphasizing the competencies required, determining and relating the main team values to the present task, and creating a clear team vision of project outcomes.[142]

In the *storming* stage, the project team manager must set clear parameters for project team members, listen to the views and opinions of all team members, consider opportunities, suggest perspectives, and reflect on vision and values. The *norming* stage requires project managers to encourage and model supportive behavior and open communication, develop team spirit so team members help each other, and provide positive feedback. They must also pay close attention to the *performing* stage to facilitate group problem-solving and decision-making, and reward and acknowledge performance outcomes while ensuring positive work relationships. It is necessary to consider the impact of the project phases on team activities.[143]

How a team is built up and how it enhances efficiency and productivity is quite significant for a project's success. Therefore, the project manager must consider a wide range of aspects for effectively enhancing individual and team performance.

Team performance

Team performance refers to the extent to which the team attains its objectives.[151] It can also refer to the extent to which outcomes exceed the task requirements, conducting work collaboratively, and meeting the personal needs of team members.[152] The building of trust, clarification of roles, consideration of the influence of culture, diversification of the team, expression of appreciation, instillation of a culture of learning, prioritization of health, proper use of technology, and recognition of good performance are some of the critical factors that help project managers improve the performance, efficiency and productivity of individuals and the team.[153]

Homogeneous and heterogeneous teams

A *homogeneous team* is a group of people who have similar characteristics, like age, educational background, ethnicity, experience and gender. Members of homogeneous teams feel included due to shared languages, cultures and backgrounds, and this makes communication more effective and removes any misunderstanding among them.

A *heterogeneous team* is a group of people, each of whom has valuable information, knowledge, or skills. Their diverse characteristics, mixed ethnicities and different backgrounds enable them to collectively make better and more informed decisions.[154]

Belbin's team roles

Belbin discovered that a diverse behavioral mix is quite pivotal for successful teams. The Belbin team roles consider the different behavioral attributes required for team progress, and are explained in Table 2:

Table 2: Belbin's team roles [155]

Role	Explanation
Resource investigator	The resource investigator facilitates bringing up new and unique ideas for the team members.
Team worker	A team worker helps the team members identify and complete their respective tasks through versatility.

Role	Explanation
Coordinator	A coordinator helps the team define its objectives for delegating tasks appropriately.
Plant	A plant facilitates problem-solving in unconventional ways.
Monitor evaluator	A monitor evaluator weighs the options of team members with dispassionate and impartial judgment.
Specialist	Specialists provide in-depth and detailed information to the team members in their respective areas.
Shaper	A shaper helps maintain the focus of team members on the objectives.
Implementer	An implementer is required for building and implementing a sound strategy in the best possible way.
Completer finisher	The role of a completer finisher is to adopt quality tools for polishing the tasks and recognizing errors at the end of the project in order to deliver quality results.

Power sources

Project managers are considered the top authorities responsible for attaining project objectives. Different sources of power can help them achieve the desired outcomes, such as coercive, expert, legitimate, referent and reward power.

Coercive power is the power to generate fear among team members for the consequences of not adhering to the project compliances. Such fear is quite significant because it prevents any unethical activity by team members that would hamper project success. Expert *power* or level of expertise or skills is another source of power through which the project manager can influence team members to work in alignment with the project scope. The expertise of the manager can inspire the team members to excel in whatever job role they are assigned. *Legitimate power* reflects the formal right of the project manager to demand anything in line with the project objectives from subordinates. The *referent power* or expertise of the manager can inspire the team members to excel in whatever job role they are assigned. *Reward power* helps project managers influence and motivate team members. Rewarding the team members for their achievements can help boost their confidence, which in turn enhances their productivity and encourages them to work for project goals and objectives.[156]

Good communication, unity of direction and clear goals help project managers accomplish group tasks for organizational advantages.[157]

Different styles of group and individual management

There are various styles for managing individuals and groups, such as the directive, delegative and participative styles. The *directive style of management* helps managers manage the project team by monitoring their performance and behavior. The managers give clear instructions to the team members about work and progress. The *delegative style of management* helps project managers delegate project tasks to subordinates or team members, but, in its original form, it is not considered a viable practice because it does not allow for consultation. Instead, a little direction with minimum supervision should be provided under this style to the team members. A *participative management style* is considered the best practice for interacting with group and individual team members of the project because it welcomes the opinions of each team member and helps pool a wide range of ideas to achieve the desired outcomes.[158]

How to build a productive team

A 2021 article by Master Class[159] lists nine steps for building a productive and efficient team. Step 1 is to establish the expectations of the employees at the very beginning of the project. The second step is to respect every member of the team as an individual and thereby implement a sound team environment to encourage them to attain common goals. Step 3 is to develop and encourage connections among team members in order to direct collective actions to the attainment of team goals. The fourth step to building a productive team is to practice emotional intelligence regularly and adopt effective leadership to treat and respect individual differences. Step 5 is concerned with positively motivating the team members. The sixth step is to develop strong relationships by establishing a sound communication system. Step 7 is to ensure that the team members remain motivated by rewarding them for their good work and dedication throughout the project. The eighth step is to diversify the team. Step 9 is to establish a team that can be trusted.[159]

A project manager can also build a productive and efficient team by defining the purpose of the project or team, determining strategic and attainable goals, setting expectations, implementing effective review and monitoring procedures, and rewarding and celebrating the efforts of the team members.[160] Much of this is done at team meetings.

Project team meetings

Good management of project team meetings is crucial for making informed decisions together. Such meetings are fruitful for assigning team members their roles and responsibilities and establishing clear procedures for their respective work. The use of technology, planning meetings, setting the meeting purpose, circulating the agenda, recording brainstorming results, and documentation are some of the critical factors that help carry out meetings effectively.

The American Management Association[161] lists setting ground rules, encouraging participation, seeking alternatives, clarifying issues, discussing consequences, and summarizing discussions as some of the proper techniques for effective management of meetings. There are kick-off, walk-through, planning, problem-solving, debriefing, milestone and presentation meetings. The *kick-off* meeting is used to introduce participants to the meeting and project goals. *Walk-through* meetings clarify information through the project plan. *Problem-solving* meetings help derive alternate solutions for project problems. *Debriefing* provides critical information to the top level of management. It focuses on approvals, budget, goals, scheduling and other vital aspects. *Milestone* meetings focus on monitoring the progress of the project and taking appropriate measures for its timely completion. *Presentations* help in reflecting the final project outcomes to stakeholders and project managers. These are all part of good communication.

Communication and information

Communication refers to exchanging thoughts, ideas and beliefs.[162] It is pivotal for project managers to communicate with the project team members effectively and regularly to attain project objectives successfully. Information can be exchanged with the team members at meetings so that everyone can make informed decisions. The involvement of all key stakeholders associated with the project is also crucial for sharing information. Therefore, communication is of the utmost significance for keeping all parties updated and motivated about the project.[163]

Motivation and stimulation

Motivation is crucial for project success because it enhances the team members' productivity so they can attain the desired outcomes.[164] Setting intelligent objectives, directing team members, conducting regular reviews of the project progress, rewarding team members for their achievements, restraining punishment, and encouraging interaction among teammates are some of the best practices that project managers can employ for motivating team members.[165] Positive motivation helps them stimulate the team members to do their best to achieve better outcomes. The motivation factor also reduces the chances of project failure to the maximum possible extent.

Good team management is the key driver or contributor in helping employees become professional business managers and enabling them to achieve their personal goals while working on common project goals and organizational objectives. Constant guidance and support by the project manager are needed for the effective teamwork that helps team members unlock their potential and sharpen their expertise. Some of the other major benefits of good team management in the context of projects and conflict resolution are a rise in people's productivity and creativity, increased work efficacy, new perspectives, increased employee satisfaction, trust-building and the low possibility of disputes or conflicts.[166] Effective team development and membership result in high-quality information exchange and trust relations among team members that contribute to robust team decision-making, proper project control, high project performance and strong individual performance. However, in the present dynamic and complex project environment, project managers' team management is exposed to diverse challenges that require constant conceptualization and increasingly scientific tactics.[167] Project managers have to keep developing more competencies.

COMPETENCE TO MANAGE THE PROJECT TEAM

Experience, training and skills make project managers competent. The core competence of a project manager is the ability to carry out project activities so as to attain the desired outcomes to the defined standards.[2] A competent project manager should have a wide range of soft skills, such as communication skills for effective communication with team members and stakeholders, leadership skills for better management of work, self-awareness, confidence for boosting the morale of team members, and stakeholder engagement for delivering quality. A competent project manager can also motivate and support team members, enabling them to carry out the work within the defined scope.[168] Therefore, project managers' competence has considerable significance because it assists in delivering essential project requirements to the clients timeously and satisfactorily, and reduces or eliminates the chances of project failure.

Some of the key benefits of competent project management are less conflict, quick resolution of disputes, and increased risk resilience, with better team communication, better trust and understanding between different team members, and constructive feedback for improvement in the performance of individual team members. Additionally, competent team management by project managers helps individual employees to work on their weak areas and polish their strengths or skills, thereby fostering their confidence and teamwork satisfaction level.[169]

Project managers often face the issue of the disengagement of team members who feel directionless or have not been given proper direction. Therefore, project managers need to pay attention to providing clarity about team goals and project outcomes to team members to decrease their feelings of disengagement.[157]

Overall, secondary sources agree that project team management is a critical task, and it needs leadership competencies wherein conflicts between team members and different work styles are some of the biggest challenges that directly decrease the efficacy and productivity of team members.

Technical and professional competencies

A project manager is required to have several forms of technical and professional competence in order to ensure the success of a project. For instance, effective time management and communication competencies are required to meet the deadlines of the project and ensure that expectations related to quality are met.[170] Organizational awareness is a key professional competency required for a project manager to have a proper sense of the chain of command and responsibilities.[171] It has been found that a project manager also requires problem-solving competency to overcome difficulties and complexities during the project.[172]" Additionally, project managers require effective leadership competency to make sure that the desired progress of the project is achieved.[171]

Udo and Koppensteiner[173] add to the list of the key technical and professional competencies required in project management active listening skills, adaptability, the application of industrial knowledge, aptitude, being a good role model, a can-do attitude, commitment, common sense, confidence, conflict resolution, creativity, effective mentoring, clear vision, fairness, flexibility, good negotiation skills, influential skills, insightful decision-making, lifecycle management, open-mindedness, organizational savvy, political sensitivity, proactive tactics, proper direction, sound judgment, team building traits, trustworthiness, and the use of project management techniques or tools.[173]

Project competencies that Aston[174] views as compulsory for project managers are project scheduling, project planning, project initiation, process management, documentation development, project control, task management, risk management, PM tools use, research and organization.

Among the mixed bag of key project competencies listed in an article by the Indeed Editorial Team[175] are active listening, analytical skills, breathing techniques, budget preparation, business strategies, collaboration, decision-making, decluttering the workspace, delegation, financial report preparation, giving feedback, goal setting, industry trends analysis, leadership, marketing, oral communication, planning, receiving feedback, resource allocation, strategy creation, stress management, visionary thinking and written communication.[175] Another article by the Indeed Team[176] reveals several other technical and professional competencies required by contemporary project managers to ensure the successful delivery of a project with the right quality. Those competencies include conflict management, cost management, data analysis, motivation, organization, policy knowledge, prioritization, quality management, research skills, risk management, software proficiency and, of course, project management methodologies and frameworks.[176]

Certification by the International Project Management Association

The International Project Management Association has a certification scheme (https://ipma.world/ipma-certification/) with nine competence elements divided into three categories: perspective, people and practice. These elements are presented in Table 3, along with their relevance in the role of the project manager:

Table 3: Elements of International Project Management Association certification [177]

Category	Elements	Role of Project Manager
Perspective	<ul><li>Strategy</li><li>Governance, structures and processes</li><li>Compliance, standards and regulations</li><li>Power and interest</li><li>Culture and values</li></ul>	The role of the project manager is to align strategy with the mission, exploit opportunities, and review success factors. It is the project manager's responsibility to review the process regularly and monitor whether team members are adhering to the norms or not. Project managers are also required to respect culture and values to enhance team members' confidence and encourage them to work toward project goals.

Category	Elements	Role of Project Manager
People	- Self-reflection and management - Personal reliability and integrity - Personal communication - Relationship and engagement - Leadership - Teamwork - Conflict and crisis - Resourcefulness - Negotiation	The role of the project manager is to take responsibility for attaining objectives and communicating effectively with team members about them. Project managers should use their relationship and leadership skills while dealing with team members to enhance their productivity and resolve any conflict. Effective promotion of teamwork and optimum use of resources are also critical roles of the project manager.

Category	Elements	Role of Project Manager
Practice	- Project design - Requirements, objectives and benefits - Scope - Time - Organization and information - Quality - Finance - Resources - Procurement - Plan and control - Risk and opportunity - Stakeholders - Change and transformation	Project managers play a crucial role in designing project objectives and scope, and determining the time to complete the project. They are also responsible for various other aspects, such as the quality of the project, the budget, effective use of resources, planning, involvement with stakeholders, and modifying the project as necessary. All this helps them reduce the chances of project failure.

Virtual teams course completion and evaluation

The advent of COVID-19 led to increasing demand for virtual team management in terms of gaining knowledge and measuring performance in a virtual setting. Training of a virtual team can be completed and evaluated using the gap analysis method: A gap is identified between the current performance and the standard or benchmark performance that is expected after the course completion. Hacker *et al.*[178] recognized various factors that should be evaluated in the gap analysis of virtual teams' training needs, such as technical capabilities, consequences, moderators (culture and functional diversity) and, lastly, consequences (individual and team performance, satisfaction, morale, learning effectiveness, commitment, cohesion, support and quality). In work settings since 2019, virtual team course completion is intended to manage team performance regardless of the working environment. Challenges are also considered in the course evaluation to manage virtual team knowledge development and smooth work processes while working on the projects.[179]

Online courses are also available from other institutions, including the PMI.

Project Management Institute methodologies for managing project teams

There are several PMI methodologies available to advise project managers how to approach a project team and lead them effectively toward the attainment of common goals. The PMI methodologies are used around the globe to manage projects effectively and deal with project teams in a well-defined manner. Indeed, according to a study conducted by Hubstaff[180] in 2021, around 39% of the business corporates around the world combine them into hybrid practices for project management.

Some of the popular PMI methodologies available for use alone or hybridized are agile, scrum, lean, kanban, scrumban, waterfall, PRINCE2, eXtreme Programming, and PMI's PMBOK. The PMI's PMBOK is a prominent methodology that helps a project manager apply the universal standards effectively in a manner based on the waterfall technique. According to the PMBOK methodology, the project must comply with five basic processes: initiation, planning, execution, control and closure. The methodology also provides numerous project management techniques to complete the desired project effectively.

The PMBOK methodology is complementary to PRINCE2 and merely a reference guide.[181]

PRINCE2 methodologies for managing project teams

The PRINCE2 methodology provides seven different principles or steps to successfully complete a project: starting up, directing, initiating, controlling every stage, managing product delivery, managing state boundaries, and closure.

It is in the start-up phase of the PRINCE2 methodology that the project team is designed and appointed, and the team structure is defined.[182] Then the project manager has to motivate the team members and guide them toward the specific project goals and objectives.

SITUATIONAL MANAGEMENT AND LEADERSHIP

Situational leadership and management refer to the implementation of an effective management style that can adjust to the requirements of tasks and situations so as to ensure that the needs of project team members are met. The predominant situational leadership styles used by project managers around the world are *telling* and *participating* because they are highly effective ways to lead a project team for organizational or project success.[183]

Teams in the organizational context

Teams play a vital role in an organization because the productivity and progress of any task and project depend upon the performance of a team. In a broader sense, a team can be regarded as a group of different people with specified performance goals who are committed to the attainment of a common purpose. This means a team must be effectively manageable in terms of size and skills, and dedicated to attaining collective team goals in an organizational context.[184]

Managing the project team

The management of the project team is crucial for the successful delivery of a project using the allocated and available organizational resources. In fact, effective management of the project team helps the project manager to enhance productivity and efficiency and thereby secure the specified project delivery. Managing the project team effectively ensures cooperation and coordination among team members, which in turn helps overcome cooperative tensions and conflicts during a project.[185]

Aligning organizational goals and project goals

Strategic alignment of organizational goals and project aims is vital for increasing project success rates in organizations because poor strategic alignment between project and organizational goals not only leads to poor project performance, but may also lead to project failure.[157] Projects that are not properly aligned with the strategy of the organization create challenges for both the project manager and team members in terms of quantifiable and clear benefits. Strategic alignment helps in eliminating waste, securing the project management objectives, improving resource allocation decisions, and increasing the focus of project team members, in addition to enabling business enterprises to leverage benefits from the completion of projects. Similarly, strategic alignment and clear goals enhance team members' understanding of outcomes and performance expectations, so it is useful for project managers to focus equally on strategically aligning project goals with organizational strategy and on project team management.[6]

The importance of teamwork has increased in today's virtual and globalized work environment where staffing is diverse. A collaborative approach in organizational projects enables different team members and departments to work in unison and complement each other by maximizing strengths and reducing weaknesses.[186] Fulfilling the goals of organizational projects requires interdependency and cooperation between varied departments and team members. On the other hand, in cases of misalignment, confusion and poor coordination between team members and departments working on the same project adversely affect the final outcomes and deliverables in terms of quality, cost and time.

Project goals can only be effectively attained when project managers proactively engage in improving teamwork and determining the teamwork issues faced while working on a common project through innovative planning and team management.[167]

Delegating and transmitting information

Delegation is a crucial function of team management to be performed during the lifecycle of a project to educate the project team about its tasks, responsibilities, authorities, and even chain of command. *Delegation* refers to the assignment of responsibility and authority to perform a task to another person, but full information about the desired task must be clearly transmitted to enable the delegate to carry it out.

Risks of team management

Litvinsky[187] viewed the key risks associated with team management as procrastination, whining, working-to-rule, passive or aggressive behavior, and toxic clients. An actiTIME blog post[188] perceived them as decreased performance, difficult employees, employee burnout, poor teamwork, and unrealistic expectations.

Team mood and conflict resolution

Projects fail not just because of poor project planning and execution but also because of poor team relationships. Maintaining a sound relationship between team members is crucial for the project manager to ensure that productivity and coordination during the course of project development remain progressive and intact. The manager must be able to handle their mood and workplace conflicts because these have direct effects on engagement, motivation and team cohesion.[189] Team members' moods impede collaboration and coordination because their participation in the team projects varies as per the mood.

Managing the team's mood is essential for the employees' positive behavior and work attitude; otherwise, anxiety, conflict issues, distress and work stress could arise and ultimately lead to the project's failure. Furthermore, the moodiness of individual members in the team project negatively affects creativity.[190] The project manager must be able to determine the cause of the mood.

Unfortunately, the angry mood of a team member is sometimes due to the manager's rude behavior and unfavorable practices; they cause negative emotions, frustration and work conflicts.[191]

A positive and encouraging culture with influential and transformative leadership is essential for sustaining a positive mood and avoiding mood swings and conflict situations. If the team members are in a good mood and their perceptions are aligned, they are more likely to understand the intended project goal and make individual suggestions for working while carrying out the project activities.

However, according to Tabassi, Abdullah and Bryde,[110] conflict in most current projects has escalated due to cultural differences or differences in the team members' perceptions that raise coordination issues.

Multicultural teams and diversity management

Project execution with a diverse and multicultural team makes it easier to attain project goals with the support of the people's different experiences, knowledge and motivations.[192] Diversity management is therefore in high demand for knowledge creation in the contemporary work environment.

However, it is not easy to manage a multicultural team without any challenges because their differences raise team conflicts and hurdles in team cohesion — and a high probability of failure. Different work values, beliefs, perceptions and ways of doing work must be accommodated to ensure successful project achievement. Sogancilar and Ors[193] found that the misinterpretation of ideas or opinions is one of the major issues in multicultural team management due to ineffective communication and coordination.

Poor communication because of cultural complexity and language barriers is a big issue in the construction industry because it leads to poor performance by multicultural teams. Therefore, as Ochieng and Price[194] pointed out, when managing a multicultural team, the project manager should play a key role in clearly communicating the defined roles and responsibilities, along with norms for the entire team to become cross-culturally competent.

Although one of the pressing issues faced by managers is multiculturalism or multicultural teams, consideration of other factors that can help in successful project execution and completion is effective in a multicultural environment too.[186] Some of the many factors that influence project success and create challenges for project managers while executing team projects are: access to information, communication skills, emotional attitude, a desire to learn (or not), place and power dynamics in a group, professional qualifications, and risk-taking propensity, as well as social and general knowledge. Efficacious communication channels, goal orientation, good interpersonal relations, the development of mutual trust and an understanding of cultural aspects are other prominent factors that shape and affect project team management in organizations.

CONCLUSION

Although our understanding of the importance of consciously working to prevent project failure on large projects has increased, the underlying reasons for failure remain an issue and sometimes a point of contention for project management practitioners and academics alike. Without a doubt, there is still a lot to learn from studying project failure. As previously specified, project management is intrinsically tied to the time, cost and quality paradigm, and projects that are challenged are typically forced to make trade-offs in time and budget estimates, and even in quality. Such trade-offs usually lead to escalation in which key personnel are turned against each other. There is little evidence that the issues of project failure outlined in this book have been fully addressed. My research into project failure presented here requires recognition of the influence that multiple stakeholders have on projects, and a broader-based view of leadership and stakeholder management. Development of an alternative methodology for project management, based on good leadership, stakeholder engagement and risk assessment, would lead to a better understanding of management issues, which might contribute to the more frequent delivery of successful projects.

LIST OF REFERENCES

1. Nieto-Rodriguez, A., & Vargas, R. V. (2023, February 2). *How AI will transform project management.* Harvard Business Review. https://hbr.org/2023/02/how-ai-will-transform-project-management

2. Project Management Institute. (2017). *Project manager competency development framework* (3rd ed.). Project Management Institute.

3. Prieto, B. (2015). *Theory of management of large complex projects.* LULU COM.

4. Pinto, J. K., & Winch, G. (2016). The unsettling of "settled science:" The past and future of the management of projects. *International Journal of Project Management, 34*(2), 237–245.

5. Meredith, J. R., Shafer, S. M., & Mantel Jr, S. J. (2017). *Project management: A managerial approach* (10th ed.). Wiley. https://www.wiley.com/en-us/Project+Management%3A+A+Managerial+Approach%2C+11th+Edition-p-9781119803836

6. Pryke, S., & Smyth, H. (Eds.). (2012). *The management of complex projects: A relationship approach* (2nd ed.). Wiley-Blackwell.

7. Yang, J., Shen, G., Ho, M., Drew, D., & Chan, A. (2009). Exploring critical success factors for stakeholder management in construction projects. *Journal of Civil Engineering and Management, 15*(4), 337–348. https://doi.org/10.3846/1392-3730.2009.15.337-348

8. Rezvani, A., & Khosravi, P. (2019). Identification of failure factors in large scale complex projects: An integrative framework and review of emerging themes. *International Journal of Project Organisation and Management, 11*(1), 1–21.

9. Cicmil, S., Williams, T., Thomas, J., & Hodgson, D. (2006). Rethinking project management: Researching the actuality of projects. *International Journal of Project Management, 24*(8), 675–686. https://doi.org/10.1016/j.ijproman.2006.08.006

10. Remington, K., & Pollack, J. (2016). *Tools for complex projects*. Routledge.

11. Špundak, M. (2014). Mixed agile/traditional project management methodology: Reality or illusion? *Procedia - Social and Behavioral Sciences, 119*, 939–948. https://doi.org/10.1016/j.sbspro.2014.03.105

12. Koskela, L., & Howell, G. (2002). The underlying theory of project management is obsolete. *Proceedings of the 2002 PMI Conference*, 293–302.

13. Olohan, M., & Davitti, E. (2017). Dynamics of trusting in translation project management: Leaps of faith and balancing acts. *Journal of Contemporary Ethnography, 46*(4), 391–416. https://doi.org/10.1177/0891241615603449

14. Walker, A. (2015). *Project management in construction* (6th ed.). John Wiley & Sons.

15. Kerzner, H. (2017). *Project management case studies* (15th ed.). Wiley. https://doi.org/10.1002/9781119389040

16. Crawford, J. K. (2021). *Project management maturity model* (4th ed.). Auerbach Publications. https://doi.org/10.1201/9781003129523

17. Dietrich, P., Eskerod, P., Dalcher, D., & Sandhawalia, B. (2010). The dynamics of collaboration in multipartner projects. *Project Management Journal, 41*(4), 59–78. https://doi.org/10.1002/pmj.20194

18. Pich, M. T., Loch, C. H., & De Meyer, A. (2002). On uncertainty, ambiguity, and complexity in project management. *Management Science, 48*(8), 1008–1023.

19. Davies, A. (2019). *Project management for large, complex projects* (Collaborative Research, p. 24) [Expert]. Association for Project Management. https://www.apm.org.uk/media/32612/digital_edinburgh-tram-report.pdf

20. Girginov, V. (Ed.). (2013). *Handbook of the London 2012 Olympic and Paralympic Games: Vol. Volume One: Making the Games.* Routledge. https://doi.org/10.4324/9780203126486

21. Davies, A., MacAulay, S., DeBarro, T., & Thurston, M. (2014). Making innovation happen in a megaproject: London's Crossrail suburban railway system. *Project Management Journal, 45*(6), 25–37. https://doi.org/10.1002/pmj.21461

22. Sanchez, O. P., Terlizzi, M. A., & de Moraes, H. R. de O. C. (2017). Cost and time project management success factors for information systems development projects. *International Journal of Project Management, 35*(8), 1608–1626. https://doi.org/10.1016/j.ijproman.2017.09.007

23. Lu, Y., Luo, L., Wang, H., Le, Y., & Shi, Q. (2015). Measurement model of project complexity for large-scale projects from task and organization perspective. *International Journal of Project Management, 33*(3), 610-622. https://doi.org/10.1016/j.ijproman.2014.12.005

24. Heagney, J. (2016). *Fundamentals of project management* (5th ed.). AMACOM.

25. Gomes, J., & Romão, M. (2016). Improving project success: A case study using benefits and project management. *Procedia Computer Science, 100,* 489–497. https://doi.org/10.1016/j.procs.2016.09.187

26. Wysocki, R. K. (2011). *Effective project management: Traditional, Agile, eXtreme* (6th ed.). Wiley. https://www.oreilly.com/library/view/effective-project-management/9781118016190/

27. Horwath, J. A. (2012). How do we manage? Project management in libraries: an investigation. *Partnership: The Canadian Journal of Library and Information Practice and Research, 7*(1), Article 1. https://doi.org/10.21083/partnership.v7i1.1802

28. Lalmi, A., Fernandes, G., & Souad, S. B. (2021). A conceptual hybrid project management model for construc-

tion projects. *Procedia Computer Science, 181,* 921–930. https://doi.org/10.1016/j.procs.2021.01.248

29. Sage, D., Dainty, A., & Brookes, N. (2014). A critical argument in favor of theoretical pluralism: Project failure and the many and varied limitations of project management. *International Journal of Project Management, 32*(4), 544–555. https://doi.org/10.1016/j.ijproman.2013.08.005

30. Charrett, D. (2018). Managing time, cost and quality: A tale of two buildings 1. In *The Application of Contracts in Engineering and Construction Projects* (pp. 165–179). Informa Law, Routledge.

31. Davies, A., & Mackenzie, I. (2014). Project complexity and systems integration: Constructing the London 2012 Olympics and Paralympics Games. *International Journal of Project Management, 32*(5), 773–790. https://doi.org/10.1016/j.ijproman.2013.10.004

32. Park, A., & Chang, C.-Y. (2013). Impacts of construction events on the project equity value of the Channel Tunnel project. *Construction Management and Economics, 31*(3), 223–237.

33. Joslin, R., & Müller, R. (2015). Relationships between a project management methodology and project success in different project governance contexts. *International Journal of Project Management, 33*(6), 1377–1392. https://doi.org/10.1016/j.ijproman.2015.03.005

34. Sauser, B., Reilly, R., & Shenhar, A. (2009). Why projects fail? How contingency theory can provide new insights: A comparative analysis of NASA's Mars

Climate Orbiter loss. *International Journal of Project Management, 27*(7), 665–679. https://doi.org/10.1016/j.ijproman.2009.01.004

35. San Cristóbal, J., Carral, L., Díaz, E., Fraguela, J., & Iglesias, G. (2018). Complexity and project management: A general overview. *Complexity, 2018*(October), 1–10. https://doi.org/10.1155/2018/4891286

36. Rasnacis, A., & Berzisa, S. (2017). Method for adaptation and implementation of agile project management methodology. *Procedia Computer Science, 104*, 43–50. https://doi.org/10.1016/j.procs.2017.01.055

37. Association for Project Management. (2020). *What is project management?* APM Association for Project Management. https://www.apm.org.uk/resources/what-is-project-management/

38. Akhmetshin, E., Petr, R., R.R., Z., E., Z., Aleshko, R., & L., M. (2019). Modern approaches to innovative project management in entrepreneurship education: A review of methods and applications in education. Journal of Entrepreneurship Education, 22. *Journal of Entrepreneurship Education, 22*(April), 1–15.

39. Nagarkatte, U. P., & Oley, N. (2017). *Theory of constraints: Creative problem solving.* Productivity Press, Taylor & Francis. https://doi.org/10.4324/9781315165554

40. Olawale, Y. A., & Sun, M. (2010). Cost and time control of construction projects: Inhibiting factors and mitigating measures in practice. *Construction Management and Economics, 28*(5), 509–526. https://doi.org/10.1080/01446191003674519

41. Mazlum, M., & Güneri, A. (2015). CPM, PERT and project management with fuzzy logic technique and implementation on a business. *Procedia - Social and Behavioral Sciences, 210*, 348–357. https://doi.org/10.1016/j.sbspro.2015.11.378

42. Tezel, A., & Aziz, Z. (2017). Visual management in highways construction and maintenance in England. *Engineering, Construction and Architectural Management, 24*, 486–513. https://doi.org/10.1108/ECAM-02-2016-0052

43. Davies, A., Dodgson, M., & Gann, D. (2016). Dynamic capabilities in complex projects: The case of London Heathrow Terminal 5. *Project Management Journal, 47*(2). https://doi.org/10.1002/pmj.21574

44. PRINCE2. (2021). *PRINCE2 methodology*. PRINCE2.Com. https://www.prince2.com/zar/prince2-methodology

45. Takeuchi, H., & Nonaka, I. (1986, January 1). The new new product development game. *Harvard Business Review, January*. https://hbr.org/1986/01/the-new-new-product-development-game

46. Sutherland, J., & Schwaber, K. (1995). SCRUM development process. *Proceedings of the Tenth Annual Conference on Object-Oriented Programming Systems, Languages, and Applications, Austin, Texas, USA, October 15-19, 1995*. Object-Oriented Programming, Systems, Languages and Applications Conference, Austin, TX.

47. Guévin, M. (2017, June 7). *Collaborative project management software*. Nutcache. https://www.nutcache.com/

48. Salimi, S. (n.d.). *The origins of scrum*. Agile Academy. https://www.agile-academy.com/en/foundations/the-origins-of-scrum/

49. Thesing, T., Feldmann, C., & Burchardt, M. (2021). Agile versus waterfall project management: Decision model for selecting the appropriate approach to a project. *Procedia Computer Science, 181*, 746–756. https://doi.org/10.1016/j.procs.2021.01.227

50. Project Management Institute. (2021). *A guide to the Project Management Body of Knowledge (PMBOK® Guide) and the Standard for Project Management* (7th ed.). Project Management Institute. https://www.pmi.org/pmbok-guide-standards/foundational/pmbok

51. Sutrisna, M., Ramanayaka, C. D. D., & Goulding, J. S. (2018). Developing work breakdown structure matrix for managing offsite construction projects. *Architectural Engineering and Design Management, 14*(5), 381–397. https://doi.org/10.1080/17452007.2018.1477728

52. Doval, E. (2019). Risk management process in projects. *Review of General Management, 30*(2), 97–113.

53. Hessami, A. G. (Ed.). (2019). *Perspectives on risk, assessment and management paradigms*. Books on Demand. https://doi.org/10.5772/intechopen.77127

54. Singh, H., & Kaur, C. D. (2021). *Sustainable green development and manufacturing performance through modern production techniques*. CRC Press.

55. Elena, P. (2020). *Academic research in multidisciplinary innovation*. International Science Group.

56. Bahamid, R. A., & Doh, S. I. (2017). A review of risk management process in construction projects of developing countries. *IOP Conference Series, 271*, 012042. https://doi.org/10.1088/1757-899X/271/1/012042

57. Petrovic, D. (2017). *Risk management in construction projects* [Royal Institute of Technology, Department of Real Estate and Construction Management]. https://www.diva-portal.org/smash/get/diva2:1150372/FULLTEXT01.pdf

58. Winch, G. M. (2012). *Managing construction projects: An information processing approach* (2nd ed.). John Wiley & Sons.

59. Smith, N. J., Merna, T., & Jobling, P. (2014). *Managing risk in construction projects* (3rd ed.). Wiley-Blackwell. https://www.wiley.com/en-us/Managing+Risk+in+Construction+Projects%2C+3rd+Edition-p-9781118347232

60. Sharma, S., & Gupta, A. (2020). Identification and management of risk in building and infrastructure projects. *Journal of Construction Engineering Technology & Management, 10*(2), 1–9. https://doi.org/10.37591/jocetm.v10i2.3951

61. Carlos, T. (2021). *Reasons why projects fail*. Project Smart. https://www.projectsmart.co.uk/recommended-reads/reasons-why-projects-fail.php

62. Czahajda, R. (2019). Project management practices in Polish nonprofit organisations. In Z. Wilimowska, L. Borzemski, & J. Świątek (Eds.), *Information Systems Architecture and Technology: Proceedings of 39th International Conference on Information Systems Architecture*

and Technology – ISAT 2018 (pp. 358–370). Springer International Publishing. https://doi.org/10.1007/978-3-319-99993-7_32

63. Nguyen, L. H., & Watanabe, T. (2017). The impact of project organizational culture on the performance of construction projects. *Sustainability, 9*(5), Article 5. https://doi.org/10.3390/su9050781

64. Bloch, M., Blumberg, S., & Laartz, J. (2012, October 1). *Delivering large-scale IT projects on time, on budget, and on value*. McKinsey Digital. https://www.mckinsey.com/capabilities/mckinsey-digital/our-insights/delivering-large-scale-it-projects-on-time-on-budget-and-on-value

65. Loukis, E., & Charalabidis, Y. (2011). Why do egovernment projects fail? Risk factors of large information systems projects in the Greek public sector. *International Journal of Electronic Government Research, 7*(2), 59–77. https://doi.org/10.4018/jegr.2011040104

66. Srinivas, K. (2019). Process of risk management. In A. G. Hessami (Ed.), *Perspectives on Risk, Assessment and Management Paradigms* (p. Chapter 11). IntechOpen. https://www.intechopen.com/chapters/64630

67. Odimabo, O. (2016). *Risk management system to guide building construction projects in developing countries: A case study of Nigeria* [University of Wolverhampton]. https://wlv.openrepository.com/handle/2436/618537

68. Foerster, M., & Arnold, C. (2019, July 21). *Preparing future-ready professionals: Eight steps to establish a firm risk management program*. IFAC International Federation of

Accountants. https://www.ifac.org/knowledge-gateway/preparing-future-ready-professionals/discussion/eight-steps-establish-firm-risk-management-program

69. Van Heerden, J. P. (2013). *Reducing risks in large scale projects: Investigating the integration of systems engineering principles into project management* [Stellenbosch University]. http://hdl.handle.net/10019.1/80151

70. Befrouei, M. A. R., & Taghipour, M. (2015). Identification and management of risks in construction projects. *American Journal of Civil Engineering, 3*(5), Article 5. https://doi.org/10.11648/j.ajce.20150305.15

71. Banaitiene, N., & Banaitis, A. (2012). Risk management in construction projects. In N. Banaitiene (Ed.), *Risk Management—Current Issues and Challenges* (pp. 429–448). InTech. https://doi.org/10.5772/51460

72. Kishk, M., & Ukaga, C. (2008). *The impact of effective risk management on project success.* University of the West of England.

73. Lee, E., Park, Y., & Shin, J. G. (2009). Large engineering project risk management using a Bayesian belief network. *Expert Systems with Applications, 36*(3, Part 2), 5880–5887. https://doi.org/10.1016/j.eswa.2008.07.057

74. Mhetre, K. V., Konnur, B., & Landage, A. B. (2016). Risk management in construction industry. *International Journal of Engineering Research, 5*(1), 153–155. https://doi.org/10.17950/ijer/v5i1/035

75. Blazquez, R., Martin, J. M., Mera, E., Christensen, D., Peineke, C., Graf, K., Lyon, D., & Stevens, G. (2011). *Risk quantification and risk management in renewable*

energy projects (p. 150) [Commissioned by the IRS - Renewable Energy Technology Deployment]. Altran GmbH & Co KG. http://iea-retd.org/wp-content/uploads/2011/11/RISK-IEA-RETD-2011-6.pdf

76. Szymański, P. (2017). Risk management in construction projects. *Procedia Engineering, 208*, 174–182. https://doi.org/10.1016/j.proeng.2017.11.036

77. Karimiazeri, A., Mousavi, N., Mousavi, S. F., & Hosseini, S. (2011). Risk assessment model selection in construction industry. *Expert Systems with Applications, 38*, 9105–9111.

78. Barlow, G., Woolley, P., Dew, C., & Dempsey, H. (2014). *Project risk management.* kpmg.com/nz. https://assets.kpmg/content/dam/kpmg/pdf/2014/02/KPMG-PALS-9-Project-risk-management.pdf

79. Murray-Webster, R., & Dalcher, D. (2019). *APM body of knowledge* (7th ed.). Association for Project Management. https://www.research.lancs.ac.uk/portal/en/publications/apm-body-of-knowledge(d7c02892-9762-4b71-b646-0604d419bf5d)/export.html

80. Abd Karim, S. B. B. (2015). *The development of an empirical-based framework for project risk management* [PhD, Alliance Manchester Business School, University of Manchester]. https://research.manchester.ac.uk/en/studentTheses/the-development-of-an-empirical-based-framework-for-project-risk-

81. Zhang, A., Chettupuzha, A. A., Chen, H., Wu, X., & AbouRizk, S. M. (2017). Fuzzy cognitive maps en-

abled root cause analysis in complex projects. *Applied Soft Computing, 57,* 235–249.

82. Westland, J. (2007). *The project management life cycle: A complete step-by-step methodology for initiating, planning, executing and closing a project successfully* (2nd ed.). Kogan Page Publishers.

83. Issac, G., Rajendran, C., & R.N, A. (2004). A conceptual framework for total quality management in software organizations. *Total Quality Management & Business Excellence, 15*(3), 307–344. https://doi.org/10.1080/1478336042000183398

84. American Society for Quality. (2021). *What is total quality management (TQM)?* ASQ American Society for Quality. https://asq.org/quality-resources/total-quality-management

85. American Society for Quality. (2022). *What are the benefits of total quality management? TQMcase studies.* ASQ American Society for Quality. https://asq.org/quality-resources/total-quality-management/tqm-gets-results

86. Bakar, A. H. A., Ali, K. B., & Rasheed, E. (2011). Total quality management practices in large construction companies: A case of Oman. *World Applied Sciences Journal, 15*(2), 285–296.

87. Monden, Y. (2011). *Toyota production system: An integrated approach to just-in-time* (4th ed.). Productivity Press. https://doi.org/10.1201/b11731

88. Patel, V. & Solanki, J. (2020). Just in time concept used in construction project. *International Research Journal of Engineering and Technology, 7*(6), 3298–3303.

89. Jenkins, A. (2020). *Just-in-time inventory explained: A guide*. Oracle NetSuite. https://www.netsuite.com/portal/resource/articles/inventory-management/just-in-time-inventory.shtml

90. Invest Northern Ireland. (2022). *Innovation in manufacturing: Advantages and disadvantages of just-in-time production*. https://www.nibusinessinfo.co.uk/content/advantages-and-disadvantages-just-time-production

91. Besner, C., & Hobbs, B. (2012). The paradox of risk management: A project management practice perspective. *International Journal of Managing Projects in Business, 5*(2). https://doi.org/10.1108/17538371211214923

92. Intaver Institute. (2021). *Managing risks in large projects: Project risk management software and project risk analysis software with Monte Carlo simulations by Intaver Institute*. Intaver Institute. https://intaver.com/managing-risks-in-large-projects/

93. Borek, A., Parlikad, A. K., & Woodall, P. (2011). Towards a process for total information risk management Proceedings of the 16th International Conference on Information Quality (ICIQ) 11, pp. 477–491. *Proceedings*, 477–491.

94. Serrador, P., & Pinto, J. K. (2015). Does agile work? A quantitative analysis of agile project success. *International Journal of Project Management, 33*(5), 1040–1051. https://doi.org/10.1016/j.ijproman.2015.01.006

95. Colwill, C. (2009). Human factors in information security: The insider threat – Who can you trust these days? *Information Security Technical Report, 14*(4), 186–196. https://doi.org/10.1016/j.istr.2010.04.004

96. DeLoach, J. (2016, April 1). 5 common risk management failures. *Corporate Compliance Insights.* https://www.corporatecomplianceinsights.com/5-common-risk-management-failures/

97. Keqa, A., & Dragomir, B. (2016, April). *12 reasons for risk management failure.* PECB Professional Evaluation and Certification Board. https://pecb.com/pdf/articles/90-pecb-12-reasons-for-risk-management-failure.pdf

98. Urbański, M., Haque, A., & Oino, I. (2019). The moderating role of risk management in project planning and project success: Evidence from construction businesses of Pakistan and the UK. *Engineering Management in Production and Services, 11*(1), 23–35. https://doi.org/10.2478/emj-2019-0002

99. Naoum, S. G., Harris, J., Rizzuto, J., & Egbu, C. (2020). Gender in the construction industry: Literature review and comparative survey of men's and women's perceptions in UK construction consultancies. *Journal of Management in Engineering, 36*(2), Article 2. https://doi.org/10.1061/(ASCE)ME.1943-5479.0000731

100. Teslia, I., Yehorchenkov, O., Khlevna, I., & Khlevnyi, A. (2018). Development of the concept and method of building of specified project management methodologies. *Eastern-European Journal of Enterprise Technologies, 5*(3(95)), 6–16. https://doi.org/10.15587/1729-4061.2018.142707

101. Rabechini Junior, R., & Carvalho, M. M. de. (2013). Understanding the impact of project risk management on project performance: An empirical study. *Journal of Technology Management & Innovation, 8*(Special Issue ALTEC 2011), 64–78. https://doi.org/10.4067/S0718-27242013000300006

102. Vidal, G. G., Campdesuñer, R. P., Rodríguez, A. S., & Vivar, R. M. (2017). Contingency theory to study leadership styles of small businesses owner-managers at Santo Domingo, Ecuador. *International Journal of Engineering Business Management, 9*, Article 1847979017743172. https://doi.org/10.1177/1847979017743172

103. Kreps, G. L. (2017). Diffusion theory in integrative approaches. In *Oxford Research Encyclopedia of Communication*. https://doi.org/10.1093/acrefore/9780190228613.013.251

104. Wani, T., & Ali, S. (2015). Innovation diffusion theory review and scope in the study of adoption of smartphones in India. *Journal of General Management Research, 3*(2), 101–118.

105. Almarri, K., & Gardiner, P. (2014). Application of resource-based view to project management research: Supporters and opponents. *Procedia - Social and Behavioral Sciences, 119*, 437–445. https://doi.org/10.1016/j.sbspro.2014.03.049

106. Taylor, J. (2008). *Project scheduling and cost control: Planning, monitoring and controlling the baseline*. J. Ross Publishing.

107. Aliyu, A. M. (2012). Project management using critical path method (CPM): A pragmatic study. *Global Journal of Pure and Applied Sciences, 18*(3–4), Article 3–4. https://doi.org/10.4314/gjpas.v18i3-4.11

108. Rashid, R., Hoseini, S. F., Gholamian, M. R., & Feizabadi, M. (2015). Application of queuing theory in production-inventory optimization. *Journal of Industrial Engineering International, 11*(4), 485–494. https://doi.org/10.1007/s40092-015-0115-9

109. Haghighinejad, H. A., Kharazmi, E., Hatam, N., Yousefi, S., Hesami, S. A., Danaei, M., & Askarian, M. (2016). Using queuing theory and simulation modelling to reduce waiting times in an Iranian emergency department. *International Journal of Community Based Nursing and Midwifery, 4*(1), 11–26.

110. Tabassi, A. A., Abdullah, A., & Bryde, D. (2019). Conflict management, team coordination, and performance within multicultural temporary projects: Evidence from the construction industry. *Project Management Journal, 50*(1). https://doi.org/10.1177/8756972818818257

111. Shankarmani, R., Pawar, R., Mantha, S., & Babu, V. (2012). Agile methodology adoption: Benefits and constraints. *International Journal of Computer Applications, 58.* https://doi.org/10.5120/9361-3698

112. Tytkowska, M., Werner, A., & Bach, M. (2015). Project management in the Scrum methodology. In S. Kozielski, D. Mrozek, P. Kasprowski, B. Małysiak-Mrozek, & D. Kostrzewa (Eds.), *Beyond Databases, Architectures and Structures* (Vol. 521, pp. 483–492). Springer Inter-

national Publishing. https://doi.org/10.1007/978-3-319-18422-7_43

113. Gamil, Y., & Rahman, I. A. (2017). Identification of causes and effects of poor communication in construction industry: A theoretical review. *Emerging Science Journal, 1*(4), 239–247. https://doi.org/10.28991/ijse-01121

114. Owen, T. (2019, December 10). Cursed megaprojects: Why are they never finished on time? *Pbctoday.* https://www.pbctoday.co.uk/news/planning-construction-news/why-megaprojects-fail/69006/

115. Shenhar, A. (2015). What is strategic project leadership? *Open Economics and Management Journal, 2,* 29–37. https://doi.org/10.2174/2352630001502010029

116. BBC North East & Cumbria. (2019, December 10). Newcastle University £39m project halted by construction firm collapse. *BBC News.* https://www.bbc.com/news/uk-england-humber-50731001

117. Martin, N. (2020, February 4). The UK's most over budget construction projects. *Construction Industry News.* https://cinmagazine.co.uk/the-uks-most-over-budget-construction-projects/

118. Elliott, B. (2020, January 13). *10 things company leaders need to do to stay relevant.* LinkedIn. https://www.linkedin.com/pulse/10-things-company-leaders-need-do-stay-relevant-bradley-elliott/?trk=pulse-article_more-articles_related-content-card

119. Consultancy.uk. (2020, May 27). *Most construction and engineering projects are unsuccessful.* Consultancy.Uk.

https://www.consultancy.uk/news/24677/most-construction-and-engineering-projects-are-unsuccessful

120. Liao, L., & Teo, E. (2018). Organizational change perspective on people management in BIM implementation in building projects. *Journal of Management in Engineering, 34*(3), Article 4018008. https://doi.org/10.1061/(ASCE)ME.1943-5479.0000604

121. Chen, W. T., Merrett, H. C., Lu, S. T., & Mortis, L. (2019). Analysis of key failure factors in construction partnering: A case study of Taiwan. *Sustainability, 11*(14), Article 14. https://doi.org/10.3390/su11143994

122. Ochieng, E., Price, A. D. F., & Moore, D. (2013). *Management of global construction projects*. Palgrave Macmillan. http://public.ebookcentral.proquest.com/choice/publicfullrecord.aspx?p=4008404

123. Almusaed, A., & Almssad, A. (2018). *Sustainable buildings: Interaction between a holistic conceptual act and materials properties*. IntechOpen. https://doi.org/10.5772/intechopen.69917

124. Camilleri, E., & Camilleri, E. (2016). *Project success: Critical factors and behaviours*. Taylor & Francis.

125. Soliman, E. (2017). Communication problems causing governmental projects delay: Kuwait Case Study. *International Journal of Construction Project Management, 9*(1), 1–18.

126. Edgington, T. (2020, February 12). HS2: Why do big government projects cost so much? *BBC News*. https://www.bbc.com/news/business-51443629

127. UK Construction Media. (2017). *Why are most construction projects significantly exceeding their budgets?* UK Construction Online. https://www.ukconstructionmedia.co.uk/news/construction-projects-significantly-exceeding-budgets/

128. McPartland, R. (2017, September 6). *What is the government construction strategy?* NBS. https://www.thenbs.com/knowledge/what-is-the-government-construction-strategy

129. Rodgers, J., Chong, H.-Y., & Preece, C. N. (2015). Adoption of building information modelling technology (BIM): Perspectives from Malaysian engineering consulting services firms. *Engineering Construction & Architectural Management, 22*(4), 424–445. https://doi.org/10.1108/ECAM-05-2014-0067

130. United Kingdom, Cabinet Office, Efficiency and Reform Group, and The Rt Hon Nick Hurd. (2014, January 29). *New guidance to reduce cost of construction projects.* GOV.UK. https://www.gov.uk/government/news/new-guidance-to-reduce-cost-of-construction-projects

131. United Kingdom, Department for Business and Trade and Department for Business, Energy and Industrial Strategy. (2019, July 22). *Policy paper: Construction sector deal.* GOV.UK. https://www.gov.uk/government/publications/construction-sector-deal/construction-sector-deal

132. Anderson, G. (2019, February 1). *New Technology To Revolutionise UK Construction.* Scaffmag - Scaffolding News & Jobs. https://scaffmag.com/2019/02/new-technology-revolutionise-uk-construction/

133. Lorek, S. (2022, April 6). *What is BIM (building information modeling)?* Trimble Construction. https://constructible.trimble.com/construction-industry/what-is-bim-building-information-modeling

134. Delgado-Hernandez, D., & Aspinwall, E. (2005). Improvement tools in the UK construction industry. *Construction Management & Economics, 23*(9), 965–977. https://doi.org/10.1080/01446190500204705

135. Engwall, M. (2012). PERT, Polaris, and the realities of project execution. *International Journal of Managing Projects in Business, 5*(4), 595–616. https://doi.org/10.1108/17538371211268898

136. Damoah, I., & Kumi, D. (2018). Causes of government construction projects failure in an emerging economy: Evidence from Ghana. *International Journal of Managing Projects in Business, 11*(1). https://doi.org/10.1108/IJMPB-04-2017-0042

137. United Kingdom, Health and Safety Executive. (2020). *Construction statistics in Great Britain, 2020.* HSE. https://www.hse.gov.uk/statistics/industry/construction.pdf

138. Halliday, S. (2008). *Sustainable construction.* Butterworth-Heinemann. https://doi.org/10.4324/9780080557168

139. Sternad, D. (2019). *Effective management: Developing yourself, others and organizations.* Red Globe Press.

140. Sling Team. (2021). 11 reasons why teamwork at the office is important. *Sling.* https://getsling.com/blog/importance-of-teamwork/

141. Bright, H. R. (2020). The importance of teamwork. *Culture and Performance.* https://www.brighthr.com/articles/culture-and-performance/teamwork/the-importance-of-teamwork-in-the-workplace/#:~:text=Teamwork%20is%20efficient%20work&text=And%20that's%20just%20another%20way,do%20it%20better%20and%20faster

142. Scott-Young, C., & Samson, D. (2008). Project success and project team management: Evidence from capital projects in the process industries. *Journal of Operations Management, 26*(6), 749–766. https://doi.org/10.1016/j.jom.2007.10.006

143. Mir, F., & Pinnington, A. (2014). Exploring the value of project management: Linking project management performance and project success. *International Journal of Project Management, 32*(2), 202–217. https://doi.org/10.1016/j.ijproman.2013.05.012

144. Pacific Crest Group. (n.d.). Founding principles of successful teams. *Strategic Development.* Retrieved March 15, 2022, from https://www.pcg-services.com/founding-principles-successful-teams/

145. United Kingdom, Department for Business, Innovation and Skills. (2010). *Guidelines for managing projects: How to organise plan and control projects.* BIS Department for Business, Innovation and Skills; http://www.nationalarchives.gov.uk/doc/open-government-licence/. https://assets.publishing.service.gov.uk/government/uploads/system/uploads/attachment_data/file/31979/10-1257-guidelines-for-managing-projects.pdf

146. Radford, K. J. (2013). *Individual and small group decisions*. Springer Science & Business Media. https://www.amazon.in/Individual-Small-Group-Decisions-Radford-ebook/dp/B00F5GETF6

147. Ferrell, O. C., Fraedrich, J., & Ferrell, L. (2021). *Business ethics: Ethical decision making and cases* (13th ed.). Cengage Learning.

148. Zoltan, R., & Vancea, R. (2015). Organizational work groups and work teams: Approaches and differences. *Ecoforum, 4*(1 (6)), 13–13.

149. Priestley, D. (2015, August 11). *Forming, storming, norming, performing: The stages of team formation*. Venture Team Building. https://ventureteambuilding.co.uk/forming-storming-norming-performing/#.YjBUDXp-BxPY

150. Brock, S. E., & McAliney, P. J. (2019). *Discourses on business education at the college level: On the boundaries of content and praxis*. Academic Studies Press.

151. Chiocchio, F., Kelloway, E. K., & Hobbs, B. (Eds.). (2015). *The psychology and management of project teams* (pp. ix, 534). Oxford University Press. https://doi.org/10.1093/acprof:oso/9780199861378.001.0001

152. Information Resources Management Association. (2017). Discrimination and diversity: Concepts, methodologies, tools, and applications. In *Https://services.igi-global.com/resolvedoi/resolve.aspx?doi=10.4018/978-1-5225-1933-1* (1–4). IGI Global. https://www.igi-global.com/book/discrimination-diversity-concepts-methodologies-tools/www.igi-global.

com/book/discrimination-diversity-concepts-methodologies-tools/168538

153. Young, A. (2021). How to improve your team's efficiency and productivity [Salesforce]. *The 360 Blog.* https://www.salesforce.com/blog/improve-team-efficiency-productivity/

154. Neck, H. M., Neck, C. P., & Murray, E. L. (2020). *Entrepreneurship: The practice and mindset* (2nd ed.). SAGE Publications, Inc.

155. Belbin. (2021). *The nine Belbin team roles.* Belbin. https://www.belbin.com/about/belbin-team-roles

156. Labrosse, M., & Medina, K. (2013, August 26). *5 ways project managers use power.* Project Manager. https://projectmanager.com.au/5-ways-project-managers-use-power/

157. Blomquist, T., Hällgren, M., & Söderholm, A. (2010). Project-as-practice: In search of project management research that matters. *Project Management Journal, 41*(1), 5–16.

158. Van der Heijden, B. (2011). *Managing cultural diversity.* Meyer & Meyer Verlag.

159. Wintour, A. (2021, September 13). *9 steps to building a strong and efficient team: How to build a strong team.* MasterClass. https://www.masterclass.com/articles/how-to-build-a-strong-team#how-to-build-a-strong-team-in-9-steps

160. Michael Page International. (2021). *Building an effective team.* Michael Page. https://www.michaelpage.ae/ad-

vice/management-advice/development-and-retention/building-effective-team

161. American Management Association. (2019). *Everything you need to know about running a project meeting.* https://www.amanet.org//articles/everything-you-need-to-know-about-running-a-project-meeting/

162. Maiti, S., & Sadhukhan, S. (2019). *E-commerce and business communication.* McGraw-Hill Education.

163. Information Resources Management Association. (2016). Project management: Concepts, methodologies, tools, and applications. In *Https://services.igi-global.com/resolvedoi/resolve.aspx?doi=10.4018/978-1-4666-7456-1* (1–4). IGI Global. https://www.igi-global.com/book/research-methods-concepts-methodologies-tools/www.igi-global.com/book/research-methods-concepts-methodologies-tools/118092

164. Rosenau, M. D., & Githens, G. D. (2011). *Successful project management: A step-by-step approach with practical examples* (4th ed.). Wiley. https://www.wiley.com/en-au/Successful+Project+Management%3A+A+-Step+by+Step+Approach+with+Practical+Examples%2C+4th+Edition-p-9781118276907

165. McHale, B. (2019). How to motivate your team for better performance in your project. *Project Central.* https://www.projectcentral.com/blog/motivate-project-team/

166. Pryke, S., & Smyth, H. (Eds.). (2006). *The management of complex projects: A relationship approach.* Blackwell.

167. Todorović, M. Lj., Petrović, D. Č., Mihić, M. M., Obradović, V. Lj., & Bushuyev, S. D. (2015). Project success

analysis framework: A knowledge-based approach in project management. *International Journal of Project Management, 33*(4), 772–783. https://doi.org/10.1016/j.ijproman.2014.10.009

168. Harrin, E. (2021, June 16). 10 essential project management competencies. *RGPM Rebels Guide to Project Management.* https://rebelsguidetopm.com/10-essential-project-management-competencies/

169. Yang, L.-R., Huang, C.-F., & Wu, K.-S. (2011). The association among project manager's leadership style, teamwork and project success. *International Journal of Project Management, 29*(3), 258–267. https://doi.org/10.1016/j.ijproman.2010.03.006

170. de Araújo, C. C. S., Pedron, C. D., & de Oliveira, F. Q. P. (2018). IT project manager competencies and team commitment: A new scale proposal. *Revista de Gestão e Projetos, 9*(1), 39–57.

171. Villanova University. (2021). 5 key project management skills for every project manager. *Villanova University.* https://www.villanovau.com/articles/project-management/key-project-management-skills/

172. Brill, J. M., Bishop, M. J., & Walker, A. E. (2006). The competencies and characteristics required of an effective project manager: A web-based Delphi study. *Educational Technology Research and Development, 54*(2), 115–140.

173. Udo, N., & Koppensteiner, S. (2004). *What are the core competencies of a successful project manager?* PMI® Global Congress 2004—EMEA, Prague, Czech Republic,

Newtown Square, PA. https://www.pmi.org/learning/library/core-competencies-successful-skill-manager-8426

174. Aston, B. (2022, January 2). *21 key project management skills for your resume in 2023*. The Digital Project Manager. https://thedigitalprojectmanager.com/personal/job-seeking/project-management-skills/

175. Indeed Editorial Team. (2021). *10 project management competencies*. Indeed.Com. https://www.indeed.com/career-advice/career-development/project-management-competencies

176. Indeed Editorial Team. (2021). *19 essential project management skills to master*. Indeed.Com India. https://in.indeed.com/career-advice/resumes-cover-letters/project-management-skills

177. Hartney, J. (2018, December 7). *Overview of the IPMA methodology*. ProjectEngineer. https://www.projectengineer.net/overview-of-the-ipma-methodology/

178. Hacker, J., Johnson, M., Saunders, C., & Thayer, A. (2019). Trust in virtual teams: A multidisciplinary review and integration. *Australasian Journal of Information Systems*, *23*, 1–36. https://doi.org/10.3127/ajis.v23i0.1757

179. Gallego, J. S., Ortiz-Marcos, I., & Ruiz, J. R. (2021). Main challenges during project planning when working with virtual teams. *Technological Forecasting and Social Change*, *162*, 120353. https://doi.org/10.1016/j.techfore.2020.120353

180. Cavey, C. (2020, May 29). All you need to know about hybrid project management. *Hubstaff Blog.* https://hubstaff.com/blog/hybrid-project-management/

181. Aston, B. (2021). *9 of the most popular project management methodologies made simple.* The Digital Project Manager. https://thedigitalprojectmanager.com/projects/pm-methodology/project-management-methodologies-made-simple/

182. Graham, N. (2010). *PRINCE2 for dummies.* John Wiley & Sons.

183. Henkel, T., & Bourdeau, D. (2018). A field study: An examination of managers' situational leadership styles. *Journal of Diversity Management (JDM), 13*(2), Article 2. https://doi.org/10.19030/jdm.v13i2.10218

184. V skills. (2020). Organisational context for teams. *Vskills Tutorials.* https://www.vskills.in/certification/tutorial/organisational-context-for-teams/

185. Le Roy, F., & Fernandez, A. (2015). Managing coopetitive tensions at the working-group level: The rise of the coopetitive project team. *British Journal of Management, 26*(4), 671–688. https://doi.org/10.1111/1467-8551.12095

186. Alias, Z., Zawawi, E. M. A., Yusof, K., & Aris, N. M. (2014). Determining critical success factors of project management practice: A conceptual framework. *Procedia - Social and Behavioral Sciences, 153,* 61–69. https://doi.org/10.1016/j.sbspro.2014.10.041

187. Litvinsky, A. (2018). *Team-building: The human factor and basic risks*. DataArt. https://dataart.com.ar/news/team-building-the-human-factor-and-basic-risks/

188. actiTIME. (2021, February 3). 5 common team management problems and solutions. *Project Management*. https://www.actitime.com/project-management/manage-team-efficiently

189. Zankovsky, A., & Von der Heiden, C. (2016). *Leadership with Synercube: A dynamic leadership culture for excellence*. Springer Vieweg.

190. Stephens, J., & Carmeli, A. (2016). The positive effect of expressing negative emotions on knowledge creation capability and performance of project teams. *International Journal of Project Management, 34*(5), 862–873. https://doi.org/10.1016/j.ijproman.2016.03.003

191. Meier, L. L., Gross, S., Spector, P. E., & Semmer, N. K. (2013). Relationship and task conflict at work: Interactive short-term effects on angry mood and somatic complaints. *Journal of Occupational Health Psychology, 18*(2), 144–156. https://doi.org/10.1037/a0032090

192. Brett, J., Behfar, K., & Kern, M. (2020). *Managing multicultural teams*. Routledge.

193. Sogancilar, N., & Ors, H. (2018). Understanding the challenges of multicultural team management. *Journal of Business Economics and Finance, 7*(3), 259–268. https://doi.org/10.17261/Pressacademia.2018.954

194. Ochieng, E. G., & Price, A. D. F. (2010). Managing cross-cultural communication in multicultural construction project teams: The case of Kenya and UK.

International Journal of Project Management, *28*(5), 449–460. https://doi.org/10.1016/j.ijproman.2009.08.001

195. Park, Y. S., Konge, L., & Artino, A. R. (2020). The positivism paradigm of research. *Academic Medicine: Journal of the Association of American Medical Colleges*, *95*(5), 690–694. https://doi.org/10.1097/ACM.0000000000003093

196. Turyahikayo, E. (2021). Philosophical paradigms as the bases for knowledge management research and practice. *Knowledge Management & E-Learning: An International Journal*, *13*(2), 209–224. https://doi.org/10.34105/j.kmel.2021.13.012

197. Sekaran, U., & Bougie, R. (2016). *Research methods for business: A skill building approach* (7th ed.). Wiley. https://www.wiley.com/en-us/Research+-Methods+For+Business%3A+A+Skill+Building+Approach%2C+7th+Edition-p-9781119266846

198. Alharahsheh, H., & Pius, A. (2020). A review of key paradigms: Positivism vs interpretivism. *Global Academic Journal of Humanities and Social Sciences*, *2*(3), 39–43.

199. Bell, E., Harley, B., & Bryman, A. (2018). *Business research methods* (5th ed.). Oxford University Press. https://global.oup.com/ukhe/product/business-research-methods-9780198869443

200. Cronk, B. C. (2019). *How to use SPSS®: A step-by-step guide to analysis and interpretation* (Eleventh edition). Routledge. https://doi.org/10.4324/9780429340321

201. Garcia, M. P. F. (2022). *Failing projects in the modern world: Quantitative analysis*. LIGS University.

202. Montgomery, D. C., Peck, E. A., & Vining, G. G. (2015). *Introduction to linear regression analysis* (4th ed.). John Wiley & Sons.

203. Armstrong, J. S., & Andress, J. G. (2011). Predictive validation of multivariate models: Regression vs. Trees. In J. N. Sheth (Ed.), *Multivariate Methods for Market and Survey Research* (eBook, p. Chapter 3). Marketing Classics Press. https://www.marketingclassicspress.com/books/chapter-3-predictive-validation-of-multivariate-models-regression-vs-trees/

204. Blann. (2018). *Data handling and analysis* (2nd ed.). Oxford University Press. https://global.oup.com/ukhe/product/data-handling-and-analysis-9780198812210

APPENDIX A

HOW I INVESTIGATED THE PROBLEMS IN THE UK CONSTRUCTION INDUSTRY

I decided to make finding out what caused the problems in the UK construction industry the topic of my PhD research and I started by setting the following objectives:

- To assess the current scenario of non-successful projects in the UK construction industry.

- To determine and critically analyze the crucial factors that caused the construction projects to be unsuccessful in the UK.

- To evaluate the efficacy of the measures adopted by the construction industry to eliminate the factors causing construction project failure in the UK.

- To propose strategic solutions to foster successful construction projects across the UK.

Since a scarcity of first-hand information makes it difficult for project managers to decide which measures to employ, I decided to undertake research in the form of a survey and statistical tests to address the knowledge gap. My research was based on the research methodology described in the next section.

HOW I CHOSE MY RESEARCH METHODOLOGY

After considering alternatives, I decided to use a positivist philosophy, a deductive approach, a descriptive design, a survey questionnaire, a random sampling approach and quantitative analysis methods, mainly regression analysis.

Positivism

I chose to use the philosophy of positivism in my PhD research because the positivist philosophy has the advantage of extracting a sensory- and experience-oriented knowledge base from a real-world setting. The positivist philosophy is aligned with the hypothetico-deductive model of science, so it was crucial to identify the structure and foundation of positivism through this lens to establish the foundation of the research.

A hypothetico-deductive model

The hypothetico-deductive model forms a circular process that begins with using theory from the literature to build a testable hypothesis, designing an experiment by identifying the variables for manipulating and measuring the group assignments, and conducting an empirical study based on experimentation. The findings from such a study are useful to inform the theory and contribute to the literature and thus complete the process that moves from examining the theory to identifying the a priori hypothesis, operationalizing the variables, experimenting, and testing or refining the theory.[195]

An a priori hypothesis is usually stated in quantitative terms wherein functional relationships can be drawn between the explanatory and causal factors and outcomes. Although positivism does not always rely on quantitative methods, its fundamental goal is to generate explanatory associations or causal relationships that can be used to logically control and predict the phenomenon in question. The purest view is that positivism has its roots in a system of logic.[196]

Positivists believe that knowledge can and must be developed objectively, irrespective of the values of the researcher or participants who influence its development.[196]

Objectivity and dualism

When knowledge is appropriately developed, it leads to truth and is congruent with reality and accuracy. For an appropriate development of truth, it is essential that there is an absolute separation between the research participants and the researcher. To accomplish such a separation, positivists operate in objectivity and dualism. In other words, positivists believe that the researcher and participants can be separated and emphasize engagement in research settings where the variables can be manipulated and controlled. In the purest form of positivism, the key focus of the study is to study the explanatory or causal relationships between the research variables. Consequently, positivism favors survey questionnaires and experimental designs. The findings drawn and developed through this philosophy are used to refine and test the theories that can lead to new questions and hypotheses for new studies.[196]

Thus, the positivism philosophy helped me retrieve critical sensory data from a natural, real-world setting about the factors in UK construction project failures that contributed to the projects' non-success, and about the measures to overcome them.[197]

I did not use an interpretivist philosophy because its huge flexibility offers biased interpretations that might hamper the results of the research.[198]

A deductive approach

I chose a deductive approach to obtain the clearest, most precise and to-the-point solutions to the research problems out of my original assumptions and theories.[197] The deductive approach enabled my research-centered hypotheses (listed on page 77 under the heading, *How I broke the problem down*) or assumptions to be tested to answer my research questions by determining the statistical relationship between the failure of construction projects and the relevant predictor factors.

I rejected an inductive approach because of the risk of data generalization and unnecessarily flexible findings that might compromise the results' accuracy.[197]

A descriptive design

I planned a descriptive design for the research project because it accurately describes all the relevant characteristics, suggestions, and prospects of participants in a quantitative study without reference to external influencers and biased interpretations.[197] A descriptive design also manages the validity and overall truthfulness of the final outcomes because the opinions of survey participants are straightforwardly described, with no integration of partial or biased interpretations generated by the researcher.[197]

A quantitative survey questionnaire

I compiled a quantitative questionnaire (see Appendix B) as my data collection method to retrieve valuable opinions and accurate perspectives from the survey population about the research problems, mainly the factors leading to the failure of contemporary UK construction projects. Any another data collection method, such as a focus group or interviews, would not have given me accurate insight into the problems, because these other methods cover broader data and extensive subjective information.[199]

A random sampling approach

I randomly selected 100 project managers from construction businesses based in the UK because random sampling not only gave me an authentic population with first-hand knowledge of construction project management failures and their concerning causes in the UK, but also provided bias-free findings, unlike purposive sampling.[199] I collected their email addresses from their businesses' websites and sent an explanatory letter.

A pilot study

After getting their consent, I formulated a questionnaire with thirteen closed-ended questions and ran a small pilot study with four participants to verify and validate the questionnaire as an appropriate instrument for the survey.

The actual survey

I conducted the survey with 100 participants to obtain their views of the factors that lead to the failure of construction projects. I stored their completed questionnaires securely in databases in the cloud.

Quantitative analysis

I used quantitative analysis methods, mainly regression analysis, to collect data from the survey. Quantitative analysis can assess cause-effect relationships among research-centered variables to accurately determine the factors that caused the problem that is being investigated.[200] SPSS regression analysis allowed the impact of multiple factors on construction project failure to be exactly and appropriately analyzed.[200]

The variables

The variables I chose for my quantitative regression analysis were "Influence on construction project failure", "Scheduling-oriented challenges", "Resource scarcities", and "Technological malfunctions".

HOW I BROKE THE PROBLEM DOWN

I formulated my assumptions into three clear, relevant and testable hypotheses, each with at least one *independent variable* (cause) and one dependent variable (effect):

H1(0):   Construction project failure is not associated with *scheduling-oriented challenges* in the UK construction industry.

H1(1):   Construction project failure is directly associated with *scheduling-oriented challenges* in the UK construction industry.

H2(0):   Construction project failure is not related to *resource scarcities* in the UK construction industry.

H2(1):   Construction project failure is directly related to *resource scarcities* in the UK construction industry.

H3(0):   Construction project failure has no relationship with *technological malfunctions* in the UK construction industry.

H3(1):   Construction project failure has a direct relationship with *technological malfunctions* in the UK construction industry.

Ethical considerations

While performing the research, I sent the participants a letter explaining why I wanted them to take part in my survey and promising confidentiality, with a form for them to sign as proof that their involvement in the research was voluntary and informed and that they consented to my ethical use of the information collected. I protected the anonymity of participants' identities, respected their values and ensured data security. In addition, I reviewed only credible secondary sources, avoided plagiarism and used citations in the text to give credit to those sources.[197]

Limitations

Survey data tend to be accurate but too concise to solve a broad research problem. In such a case, interviews would elaborate on the experiences of the expert population.[197] An exploratory design to dynamically acquire significant knowledge about the research subject, and an inductive approach to accumulating facts and evidence would eliminate the shortcomings of a descriptive design (highly static data) and a deductive approach (lack of flexibility).[197]

WHAT I FOUND OUT

The survey gathered inferential statistics and descriptive statistics.

Inferential statistics

My questionnaire first established the credibility of the participants' opinions by asking about their experience in the construction industry.

Experience

The survey was directed at respondents who had enough years of working experience to know what they were talking about when they provided pivotal information on the issues raised in the survey. Item 1 in the questionnaire was designed to record this.

Experience

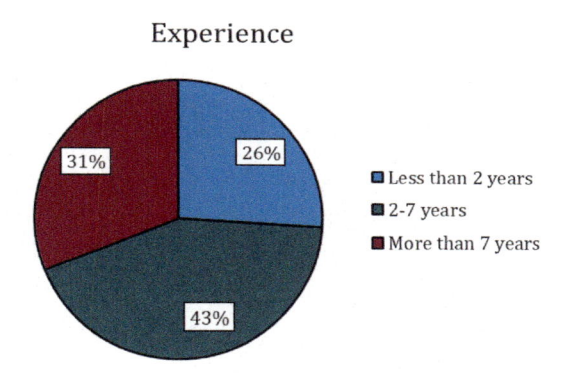

- Less than 2 years: 26%
- 2-7 years: 43%
- More than 7 years: 31%

Figure 5: Experience in the construction industry[201] (p. 90).

Answers to the first item in the questionnaire established that 26% of the 100 respondents had been working in the construction industry for less than two years; 43% had 2–7 years of experience in the industry, and 31% of the respondents had worked in the industry for more than seven years. This means that, between them, they had substantial knowledge and experience of working in construction projects.

Key motive of project management in the construction industry

The need to integrate project management into the organization as a whole has gained wider attention in construction projects, so the second item in the questionnaire was designed to establish what the respondents thought about this.

Item 2: Do you think the primary motive of project management in the construction industry is to direct and help the organization in the effective implementation of various projects and the accomplishment of its determined goals?[1]

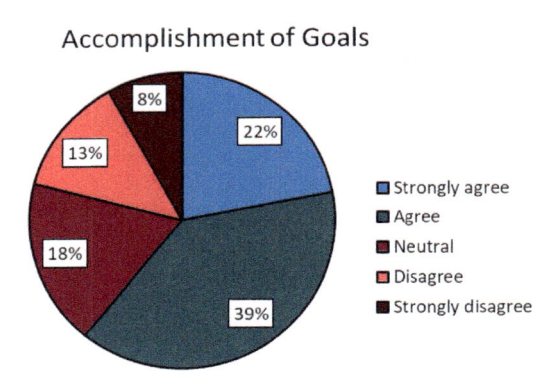

Figure 6: Project management helps fulfill agreed goals[201] (p. 92)

The responses showed that 22 of the respondents strongly agreed with the stated motive for project management, whereas 39 merely agreed, 18 were neutral, 13 disagreed and 8 strongly disagreed with the statement. Therefore, slightly more than half

---

1   The text of some of the questionnaire items has been edited for clarity but the original text is available in Appendix B.

of them agreed that the primary motive of project management in the construction industry is to direct the organization in effectively implement various projects and help it achieve its goals.

Project failures are increasing

Item 3: In your experience, has the incidence of project failure grown in the last few years?

Project Failures Are Increasing

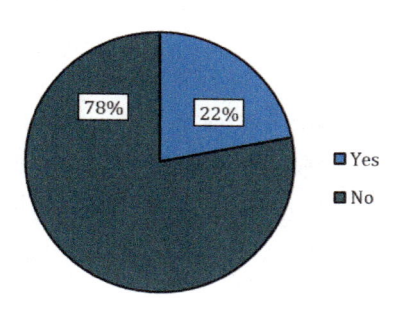

- 78% Yes

Figure 7: The incidence of project failures has increased in the last few years[201] (p. 93)

In answer to Item 3, most of the respondents showed deep concern because they had experienced a growing incidence of project failures. Only 22% of them answered "No" to the question. This highlights project failure as a key issue in the UK construction industry because it is increasing.

Critical risks in project management

Construction projects in the UK face a multitude of risks, some of them critical. Item 4 in the questionnaire was designed to find out which risk is considered the most severe.

Item 4: Which is the most critical form of risk for project management in the construction domain: financial risks, resource scarcities, schedule-management-related risks, or inadequate levels of competency and improper abilities?

Risks Related to Project Management

- Financial risk: 21%
- Resource scarcities: 28%
- Schedule-management-related risks: 25%
- Inadequate competencies and improper abilities: 26%

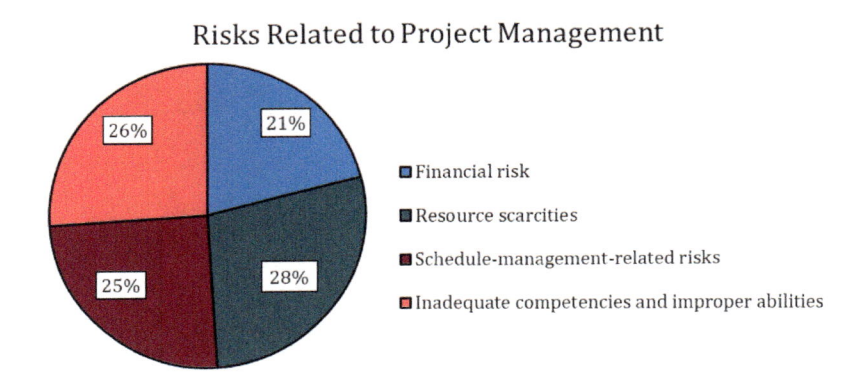

Figure 8: *Prominent risks in managing UK construction projects* [201] (p. 94)

The responses showed that 21% of the total respondents saw financial risk as the worst risk, whereas 28% emphasized resource scarcity, 25% saw schedule-management-related risks as the most severe, and 26% were most concerned about inadequate competencies and improper abilities. Analysis revealed resource-scarcity-related risk as marginally the most critical risk, followed by the risk posed by inadequate competencies and improper abilities, then schedule management-related risks, and finally financial risks.

Common scheduling challenges

Considering the risks identified in Item 4, Item 5 was designed to establish more clearly how concerned the respondents were about scheduling challenges.

Item 5: Scheduling-oriented challenges commonly confronted by project teams while undertaking any project might lead to the project's failure to be accomplished promptly.

Scheduling-oriented Risks

- Strongly agree: 59%
- Agree: 23%
- Neutral: 3%
- Disagree: 10%
- Strongly disagree: 5%

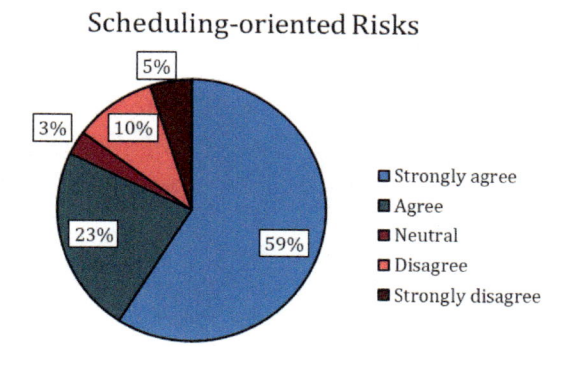

Figure 9: Scheduling-related challenges make construction projects fail[201] (p. 96)

Analysis showed that 59% of the respondents remained strongly agreed that scheduling challenges are common, 23% merely agreed, and 3% preferred to remain neutral. However, 10% of the respondents disagreed, and 5% of respondents strongly disagreed with the statement. Overall, the analysis of the distinct views of the respondents showed that a large majority of project teams were commonly confronting scheduling-oriented challenges that might result in the project's failure to be accomplished on time. This reveals an urgent need to tackle schedule-oriented challenges in construction projects effectively.

Critical path method and Gantt chart techniques

The use of these scheduling tools can considerably aid project teams in overcoming time-overrun challenges that cause project failures. In order to discover how common this knowledge is, the respondents were asked to express their opinion, from their experience and knowledge, of the relevance of Gantt charts and the critical path method to proper scheduling of the important activities in construction projects.

Item 6: The critical path method and Gantt chart techniques significantly support the optimum scheduling of critical activities linked to the project so as to complete the project in a timely manner as per the decided schedule.

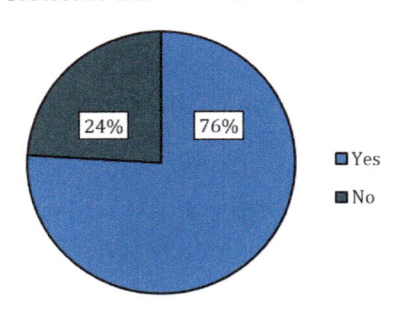

*Figure 10: **The Critical Path Technique and Gantt charts in optimal scheduling***[201] (p. 97)

More than three-quarters of the respondents marked their answer as "Yes", indicating that most of the skilled and knowledgeable people in the UK construction sector acknowledge the significant contribution of the critical path method and Gantt charts to the time-effective scheduling of vital project activities. This proves that entities that repeatedly encounter scheduling-related issues can embrace these techniques to exercise appropriate control over different tasks within the affirmed timeline, thereby overcoming time-related risks without compromising quality.

Resource scarcity or inadequate management of resources

To learn about the varied risks factors accountable for non-successful construction projects, the respondents were asked to express their level of agreement with the statement in Item 7, which specifically focused on whether resource-scarcity risks and communication risks lead to unsuccessful projects.

Item 7: The challenge of resource scarcity or inadequate management of resources also hampers successful project management.

Resource Scarcities

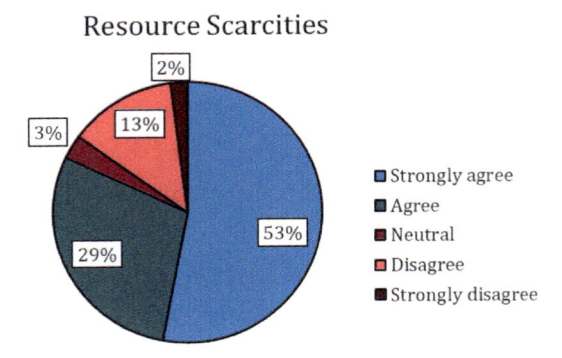

Figure 11: A scarcity of resources can result in a failed project[201] (p. 99)

The responses obtained in this context signify that the majority of the respondents believed that ineffective management of essential resources hampers the efficacious accomplishment of projects. This deduction is based on the survey responses, where 53% of the respondents indicated their strong agreement with the statement, while 29% agreed with it, and 3% of the respondents had no clue about whether resource scarcities or their mismanagement hampers successful project management since they marked "Neutral" as their response. Another 13% disagreed with the statement, and the remaining 2% strongly disagreed with it.

Factors that influence the failure of construction projects in the UK

Since I wanted to determine the foremost factor responsible for the surging incidence of project failures in the construction sector of the UK and to gain real-world evidence in this regard, I included a separate question about it in the survey questionnaire centered on such factors.

Item 8: Which factors exert the most influence on the failure of construction projects in the UK: poor management of the communication process, increasing accidents and faults, budget overrunning and schedule overrunning, skills shortage, or inadequate planning?

Factors Contributing to Project Failure

- Poor management of communication process — 13%
- Increasing accidents and faults — 26%
- Budget overruns and schedule overruns — 21%
- Skills shortage and inadequate planning — 21%
- All of the above — 19%

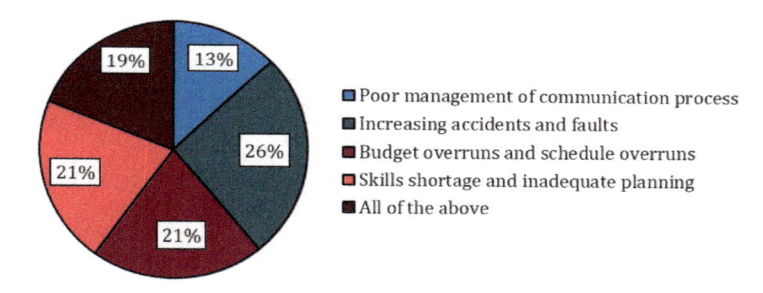

Figure 12: ***Key factors that cause the failure of construction projects in the UK***[201] (p. 100)

In their responses, 13% of the survey participants perceived poor administration of communication process as the critical factor that prevents the effective completion of construction projects. Another 26% suggested that the mounting frequency of faults and accidents substantially obstructs the efficacious accomplishment of projects, while 21% of the respondents indicated schedule overruns and budget overruns as the main obstacles to the successful completion of projects, and an equal 21% of respondents reported that a shortage of the requisite skills and inadequate planning are the principal factors causing project failures. Nearly one-fifth of the respondents (19%) marked "All of the above" as their response. Therefore, it can be inferred that all the listed factors have been observed as critical causes of failure in UK construction projects.

Technology, project objectives, and access to information

In order to dive deeper into identifying the main factors accountable for increasing non-successful construction projects, a separate statement about interruption by technological malfunctions was raised as Item 9 for the respondents' expert assessment.

Item 9: Do you feel technological malfunctioning is also a significant factor that interrupts the success of many construction projects?

Technological Malfunctions

- Strongly agree — 65%
- Agree — 15%
- Neutral — 3%
- Disagree — 7%
- Strongly disagree — 10%

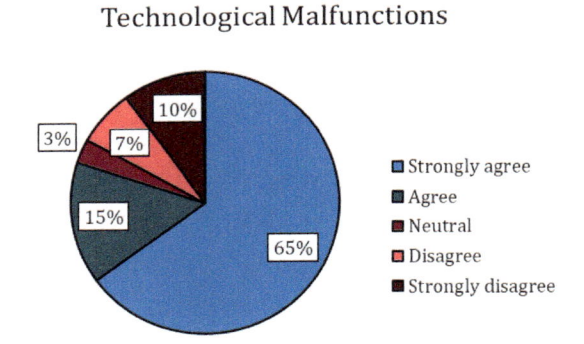

*Figure 13: **Technological malfunctioning results in project failures**[201] (p. 102)*

Sixty-five percent of the respondents strongly agreed with this statement but, while 15% agreed that it was factual, 3% of the respondents did not hold a clear viewpoint on these factors so they selected "Neutral" as their answer. Another 7% of the respondents expressed disagreement, and the remaining 10% strongly disagreed with this point. Therefore, most experts in the UK construction industry believe that the frequent malfunctioning of the technologies deployed often caused project failures.

Tools and measures

Not wanting to look only for the causes of project failure, I asked the experts what they thought the causes of success were.

Item 10: From your perspective, which of the following tools and measures substantially contribute to the effective completion of construction projects without any delays in the management of resources in line with the determined project scope: a. Matrix diagram and brainstorming, b. Critical path methodology, c. PERT (Program Evaluation and Review Technique), d. Computer-aided design software, e. All of the above

Tools that Contribute to Project Completion

- Matrix diagram and brainstorming: 18%
- Critical Path Technique: 25%
- Program Evaluation and Review Technique: 23%
- CAD designing software: 18%
- All of the above: 16%

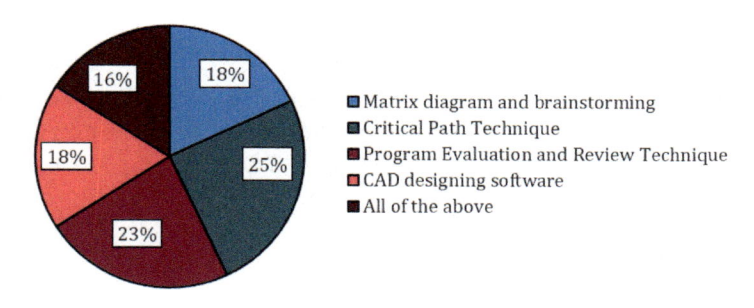

Figure 14: Measures and tools that support effective and efficient project completion [201] (p. 103)

Figure 14 shows that 16% of those who completed the survey considered brainstorming and matrix diagrams to be the key techniques to affirm the success of construction projects, while 25% of the respondents supported critical path methodology. Most of the respondents (23%) acknowledged the PERT as the most vital measure for project success, and an equal proportion of the participants affirmed the importance of CAD software in this context. Unsurprisingly, 16% of the respondents marked "All of the above" in their responses. This clearly indicates that the PERT, CAD, critical path methodology, matrix diagrams, and brainstorming are all excellent project management tools and measures that support the completion of projects on time without any reduction in quality.

AI technology

I asked the respondents to give their views on the value of adopting AI to manage operations related to construction projects and on whether AI helps reduce workload and control errors.

Item 11: The use of artificial intelligence technology in the operations of construction projects has contributed to reducing the workload, ensuring that work performed with the use of this technology is the most appropriate, and has considerably lessened the chances of mistakes.

Artificial Intelligence

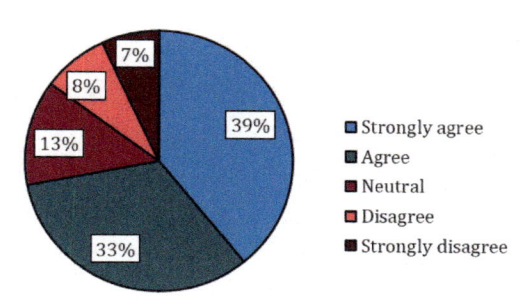

*Figure 15: **Importance of AI technology in the UK construction sector**[201] (p. 104)*

The majority of the respondents (39%) indicated strong agreement with the statement in Item 11, while 33% selected "Agree" as their response. Thirteen percent of the respondents held a "Neutral" perspective, while 15% disagreed with the significance of AI in bringing down workload and possibility of errors. Overall, analysis shows that AI in recent years is identified as an emerging phenomenon in the construction sector of the country, capable of minimizing the workload by automatically overseeing different tasks related to project management while also diminishing the probability of errors.

Agile methodology

I wanted to find out whether the project managers were familiar with the agile methodologies as measures to embrace changes and alterations and assure effective project completion.

Item 12: The agile method not only supports in effectively pursuing alterations or changes within the project during its development, but it also assures fast project management with real-time updates, low-cost investments and quality maintenance.

Agile Methodology

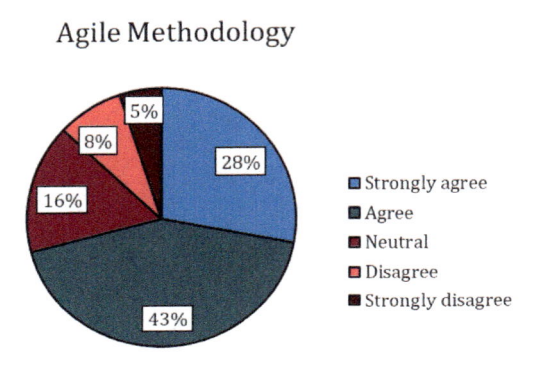

- 28% Strongly agree
- 43% Agree
- 16% Neutral
- 8% Disagree
- 5% Strongly disagree

*Figure 16: **Importance of agile methodologies in effective change and completion*** [201] (p. 106)

In response to the statement, 28% of the respondents showed strong support by marking strongly agree as their answer, while 43% of them showed that they agreed with it. Only 13% of them disagreed about the significance of agile methodology in supporting proper incorporation of changes to projects, rendering flexibility while also guaranteeing completion on scheduled time. This means that the agile method is generally accepted as a crucial project management methodology, according to the knowledge and expertise of people overseeing project management in the UK construction industry. Agile methodology is therefore an efficacious solution to the rising problem of non-successful projects in the country because it can accommodate changes and modifications in projects without causing delays.

Scrum methodology

To understand the role played by scrum technology in adding flexibility to the project management function and enabling project heads to exercise the required level of control over different project tasks, I asked the respondents to indicate their level of agreement with the statement in Item 13.

Item 13: Scrum technology used in the management of construction projects enhances the flexibility of time as well as the budget that, in turn, ensures tasks are more effectively controlled by the project owner to manage the efficacy of the project work.

Scrum Technology

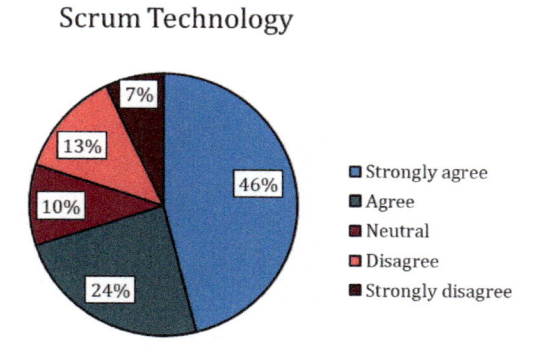

- Strongly agree: 46%
- Agree: 24%
- Neutral: 10%
- Disagree: 13%
- Strongly disagree: 7%

*Figure 17: **Scrum technology provides flexibility and effective task control**[201] (p. 107)*

A major proportion of the respondents (46%) strongly agreed that scrum technology makes a vital contribution to construction projects' efficacy, and another 24% agreed. Ten percent of the respondents displayed a neutral opinion and another 8% did not agree with this statement. The remaining 7% of the respondents strongly opposed the statement that scrum technology gives project managers flexibility and task control advantages. Therefore, it can be inferred that construction project teams should consider scrum technology as a valid approach to gain flexibility in budget and time management while also increasing control over the wide-ranging tasks of project management, thus improving the project's chances of success.

Descriptive statistics

Descriptive statistics help in analyzing and interpreting the results of the study with respect to the means and variances of the data.

Table 4: Descriptive statistics[201] (pp. 108-109)

Item	N	Minimum	Maximum	Mean	Std. deviation	Variance
Risks related to project management	100	1.00	4.00	2.5600	1.09471	1.198
Scheduling-oriented challenges	100	1.00	5.00	1.7900	1.20013	1.440
Critical path method and Gantt chart technique	100	1.00	2.00	1.2400	.42923	.184

Item	N	Minimum	Maximum	Mean	Std. deviation	Variance
Resource scarcity	100	1.00	5.00	1.8200	1.11355	1.240
Factors contributing to failure of projects	100	1.00	5.00	3.0700	1.32768	1.763
Technological malfunction	100	1.00	5.00	1.8200	1.35870	1.846
Tools contributing to completion of projects	100	1.00	5.00	2.8900	1.34010	1.796
Artificial intelligence	100	1.00	5.00	2.1100	1.21352	1.473
Agile methodology	100	1.00	5.00	2.1900	1.08892	1.186
Scrum technology	100	1.00	5.00	2.1100	1.30960	1.715
Valid N (listwise)	100					

Risks related to project management

The mean score of the variable "risks related to project management" is 2.560, which shows that most of the respondents believed that schedule-management-related risks are the most significant risks associated with project management in the construction sector. On the other hand, the standard deviation of the variable is 1.09, which shows that there was high variation in the responses of the project managers who completed the survey. It can be considered, therefore, that project managers do not agree about which risks related to project management are the highest.

Scheduling-oriented challenges

The average value of the variable "schedule-oriented challenges" is 1.79, which indicates that most of the participants agreed that schedule-oriented challenges are some of the most critical challenges in project management. The high standard deviation of the variable (1.20) shows that the views of the participants varied widely about the severity of the problem.

Technological malfunction

The average score of the "technological malfunction" variable (1.82) is close to its minimum value, which shows that most of the respondents believed that technological malfunction is a very common risk for project management.

Artificial intelligence

The mean value of the variable "artificial intelligence" is 2.11, which shows that a large number of the project managers surveyed believed that the use of AI in project management has indeed positively benefited the construction industry by reducing workload and making the entire process more efficient. However, the high standard deviation of the variable (1.21) shows that most of the responses were not aligned with its mean response, depicting high variation in their opinions of how much the tool has benefited the industry.

Agile methodology

When the mean score of the variable "agile method" is analyzed, it can be interpreted that a large number of project managers agreed that the use of agile methodologies has helped in tracking real-time information about the projects. The overall standard deviation of the variable is 1.08, which is remarkably high and shows that the overall responses of the participants were again very scattered.

Overall conclusions

The above analysis of the overall mean scores shows that all the project managers in the survey believed that, in the present scenario, project management failures have increased in the construction industry. However, the use of the critical path method and Gantt charts, conscious management of resource scarcities and other factors contributing to the failure of projects, and the use of tools that contribute to the successful completion of projects, including AI technology, scrum technology and agile methods, has helped the construction industry improve its overall effectiveness. The use of all this technology has also helped firms to reduce their overall risks and manage their overall budgets.

Regression analysis: what the statistics mean

Regression analysis helps determine the relationship between the independent and dependent variables of the study. The regression analysis method mathematically analyses the overall impact of independent variables on the dependent variable with the help of a regression equation.[202] There are many advantages of conducting a regression analysis test; for instance, it facilitates correct and valid decision-making in businesses to help firms achieve high growth and stability in the global market. Regression analysis also helps firms isolate the most important factors that affect the independent variable.[203]

I performed a multiple regression analysis to address my research question: Why have construction projects failed in the United Kingdom? The three independent variables were the risks, *resource scarcities, scheduling-oriented challenges,* and *technological malfunctions,* and the dependent variable of the analysis was *increasing project failures.* The multiple regression analysis established the effect of all the independent variables on the dependent variable to explain the effect of risks on project management in the construction sector.

(1)

where Y = dependent variable of the study
X1, X2 and X3 are independent variables of the study b1, b2 and b3 are regression coefficients of X1, X2 and X3 respectively and
C is the constant of the regression equation.

Table 5: Model summary[201] (pp. 112-113)

Model	R	R-squared value	Adjusted R-squared value	Std. error of the estimate
1	.246[a]	.061	.031	.40978

a. Predictors (Constant): Resource scarcities, Scheduling-oriented challenges, Technological malfunctions

The model summary statistics in Table 5 highlight the values of the correlation coefficients for all the data. The value of R is 0.246, which shows that there is a weak correlation between the dependent and independent variables. Also, the value of R-squared is 0.061, which shows that the total variance of the data is only 6.1%.

Table 6: Analysis of variance[201] (pp. 113-114)

ANOVA model	Sum of squares	df	Mean square	F	Sig.
Regression	1.040	3	.347	2.064	.010[b]
Residual	16.120	96	.168		
Total	17.160	99			

a. Dependent Variable: Project failures are growing

b. Predictors: (Constant), Technological malfunction, Resource Scarcity, Scheduling-oriented challenges

Table 6 portrays the output of the ANOVA test that was used to analyze the variance between the group means in the regression model. It verifies the significance of the formulated regression model, and projects the variance levels in the incorporated research items.[204] In this research, the *p*-value of the ANOVA test is 0.01, showing the high statistical significance of the regression results.

Table 7: Coefficients[201] (pp. 114-115)

Model	Unstandardized coefficients		Standardized coefficients	t	Sig.
	B	Std. error	Beta		
(Constant)	1.056	.080		13.125	.000
Resource scarcities	.072	.150	.193	.480	.022

Model	Unstandardized coefficients		Standardized coefficients	t	Sig.
	B	Std. error	Beta		
Scheduling-oriented challenges	.066	.159	.189	.411	.060
Technological malfunction	-.047	.080	-.152	-.587	.558

The results in the coefficients table (Table 7), show the regression coefficients of the independent variables and the separate *p*-values of each of the independent variables are shown in the Sig. column, which depicts their individual significance in the regression test. The beta coefficient of *scheduling-oriented challenges* is 0.066, while that of *resource scarcities* is 0.72. On the other hand, the beta coefficient of *technological malfunctions* is -0.047. Thus, the regression equation can be formulated as:

The analysis of the *p*-values of all three of the independent variables reveals that the *p*-value of *resource scarcity* is 0.022. This shows that, of all the independent variables, resource scarcity contributes the most to increasing project-related risks. Its low *p*-value (below 0.05) shows that *resource scarcities* are highly significant for project failures. In addition, the beta coefficient of *resource scarcity* is 0.72, which shows that one unit increment in the *resource scarcity* increases the risk of failure of projects by 0.72 of a unit. Thus, the null hypothesis H2(0) of the study is rejected and the alternative hypothesis is accepted.

On the other hand, the *p*-value of the independent variable *technological malfunction* is 0.558, while that of *schedule-oriented challenges* is 0.060. The *p*-values above 0.05 show that the results of the regression test are not statistically significant. Since the *p*-values of these two independent variables are not statistically significant, null hypotheses $H1(0)$ and $H3(0)$ of the study are accepted.

Therefore, the results of the tests show that, of all the three independent variables, only resource scarcities significantly influence project failures in the UK construction sector..

APPENDIX B

SURVEY QUESTIONNAIRE

1. Experience:

    a. Less than 2 years

    b. 2–7 years

    c. More than 7 years

Specific:

2. Do you think primary motive of project management in the construction sector is to direct and help organization in the effective implementation of various projects and the accomplishment of the determined goals and objectives?

    a. Strongly Agree

    b. Agree

    c. Neutral

    d. Disagree

    e. Strongly disagree

3. As per your experience, is the incidence of project failures growing in the last few years?

    a. Yes

    b. No

4. Which of the following is the most critical risk concerning project management in the construction domain?

    a. Financial risk

    b. Resource scarcity

    c. Schedule management-related risk

    d. Inadequate level of competencies and improper abilities

5. "Scheduling-oriented challenge is commonly confronted by project teams while undertaking any project that might lead to project's failure to be accomplished in a prompt manner." Kindly indicate your response in this regard.

    a. Strongly Agree

    b. Agree

    c. Neutral

    d. Disagree

    e. Strongly disagree

6. Do you feel the critical path method and Gantt chart techniques significantly support in optimum scheduling of critical activities linked to the project to complete the project in a timely manner as per the decided schedule?

   a. Yes

   b. No

7. "Apart from communication risk, the challenge of resource scarcity or inadequate management of resources also hampers the successful project management." Please mark the most appropriate answer.

   a. Strongly Agree

   b. Agree

   c. Neutral

   d. Disagree

   e. Strongly disagree

8. Which among the below-listed factors exerts the most considerable influence on the failure of construction projects in the UK?

   a. Poor management of communication process

   b. Increasing accidents and faults

   c. Budget overrunning and schedule overrunning

   d. Skills shortage and inadequate planning

   e. All of the above

9. Do you feel technological malfunctioning, unclear project objectives and information's inaccessibility are also significant factors interrupting the success of many construction projects?

   a. Strongly Agree

   b. Agree

   c. Neutral

   d. Disagree

   e. Strongly disagree

10. As per your perspective, which of the following tools and measures substantially contribute to the effective completion of construction projects without any delays in the management of resources in line with the determined project scope?

    a. Matrix diagram and brainstorming

    b. Critical path methodology

    c. PERT (Program Evaluation and Review Technique)

    d. CAD designing software

    e. All of the above

11. "The use of artificial intelligence technology in the operations of construction projects has contributed to reducing the workload, and also work performed with the use of this technology is most appropriate; hence, chances of mistake are very less."

    a. Strongly Agree

    b. Agree

c. Neutral

   d. Disagree

   e. Strongly disagree

12. "The agile method not only supports in effectively pursuing alterations or changes within the project during its development, but it also assures fast project management with real-time updates, low-cost investments and quality maintenance." Kindly mark your response in relation to the given statement.

   a. Strongly Agree

   b. Agree

   c. Neutral

   d. Disagree

   e. Strongly disagree

13. "Scrum technology used in the management of construction projects confirms flexibility of time as well as the budget that in turn, ensures tasks' controlling by project owner effectively to manage project work's efficacy."

   a. Strongly Agree

   b. Agree

   c. Neutral

   d. Disagree

   e. Strongly disagree

Printed in Great Britain
by Amazon